SWANS: THEIR BIOLOGY AND NATURAL HISTORY

Whistling swan adult

Swans:

Their Biology

and Natural History

Paul A. Johnsgard

School of Biological Sciences, University of Nebraska–Lincoln

Zea Books, Lincoln, Nebraska: 2016

Abstract

The seven species of swans of the world are an easily and universally recognized group of waterfowl, which have historically played important roles in the folklore, myths and legends in many of the world's cultures. Among the largest of all flying birds, they have also almost universally been used as symbols of royalty, grace and beauty, and largely for these reasons swans have only rarely been considered acceptable as targets for sport hunting. Swans occur on all the continents except Africa, although most species are associated with the temperate and arctic zones of North America and Eurasia. Among birds, swans are relatively long-lived species, and are also among the most strongly monogamous, having prolonged pair and family bonds that strongly influence their flocking and social behavior, and contribute to the overall high degree of human interest in them. This volume of 48,000 words describes their distributions, ecology, social behavior, and breeding biology. Included are nine distribution maps, 19 drawings, and 23 photographs by the author. There is a bibliography of nearly 700 references.

Text and illustrations copyright © 2016 Paul A. Johnsgard.

ISBN 978-1-60962-081-3 paperback
ISBN 978-1-60962-082-0 e-book

Composed in Adobe Garamond types by Paul Royster.

Zea E-Books are published by the University of Nebraska–Lincoln Libraries.

Electronic (pdf) edition available online at http://digitalcommons.unl.edu/zeabook/
Print edition available from http://www.lulu.com/spotlight/unllib

UNL does not discriminate based upon any protected status. Please see go.unl.edu/nondiscrimination

University of Nebraska–Lincoln

Contents

List of Maps 7

List of Figures 7

List of Photographs 9

Preface 11

I. Introduction to the Swans of the World 13

II. Species Accounts

    Mute Swan *Cygnus olor* (Gmelin) 1789 21

    Black Swan *Cygnus atratus* (Latham) 1790 33

    Black-necked Swan *Cygnus melancoryphus* (Molina) 1792 38

    Trumpeter Swan *Cygnus buccinator* (Richardson) 1758 43

    Whooper Swan *Cygnus cygnus* (Linnaeus) 1758 55

    Tundra Swan (Whistling Swan) *Cygnus c. columbianus* (Ord) 1815 62

    Tundra Swan (Bewick's Swan) *Cygnus columbianus bewickii* Yarrell 1830 . . 74

    Coscoroba Swan *Coscoroba coscoroba* (Molina) 1782 79

III. References 85

Whooper swan, adult wing-flapping

Maps

Map 1. Eurasian distribution of the mute swan 22

Map 2. North American distribution of the mute swan 23

Map 3. Australia and New Zealand distributions of the black swan 34

Map 4. Distribution of the black-necked swan 39

Map 5. Distribution of the trumpeter swan 44

Map 6. Distribution of the whooper swan 56

Map 7. Distribution of the North American (whistling) race of the tundra swan . . 63

Map 8. Distribution of the Eurasian (Bewick's race) of the tundra swan 75

Map 9. Distribution of the coscoroba swan 80

Figures

Whistling swan 2

Whooper swan 6

Trumpeter swan 8

Tundra swan tracheal anatomy 14

1. Mute and black swan behavior 16

2. Black and black-necked swan behavior 17

3. Trumpeter and whooper swan behavior 18

4. Whooper, whistling, Bewick's and coscoroba swan behavior 19

5. Mute swan 20

6. Black swan 35

7. Black-necked swan 40

8. Trumpeter swan 46

9. Trumpeter swan 53

10. Whooper swan 57

11. Whooper swan 58

12. Whistling (tundra) swan 65

13. Whistling (tundra) swan 71

14. Coscoroba swan 83

Trumpeter swan, adult standing

Photographs

Adult male mute swan, mild threat. 24

Adult male mute swan, preening 26

Adult male mute swan, male attack 26

Adult mute swan pair. 30

Black swan, female and cygnets 36

Black swan pair 37

Nesting black-necked swan pair 41

Black-necked swan family 41

Trumpeter swan, adult male 47

Trumpeter swan pair 49

Trumpeter swan, triumph ceremony 49

Trumpeter swan, triumph ceremony 51

Trumpeter swan copulation 51

Whooper swan post-copulatory display 59

Whooper swan, female and cygnets 60

Whistling (tundra) swan. 67

Whistling (tundra) swan. 67

Whistling (tundra) swan. 69

Bewick's and whistling swans 73

Bewick's tundra swan 77

Coscoroba swan, adult and cygnet 81

Coscoroba swan, adult pair 81

Coscoroba swan, head detail 82

Preface

It has been several years since I have written anything of special note on waterfowl, as I have for several decades instead been mesmerized into thinking and writing about cranes. I have thus missed both watching and writing about waterfowl. However, in recent years I have become increasingly interested in trumpeter swans, observing them both in their remote Rocky Mountain retreats of the Yellowstone–Grand Teton region, and in their rapidly increasing numbers throughout the central Great Plains.

Because of federal re-introduction efforts that began at South Dakota's Lacreek National Wildlife Refuge, a permanent resident population of trumpeter swans has developed over the past several decades in the central Great Plains. As a result, several hundred trumpeter swans now regularly overwinter in the relatively ice-free waters of the Nebraska Sandhills, where spring-fed creeks rarely freeze, and where large and shallow Sandhills marshes provide ideal foraging and nesting habitats from spring to fall.

In addition, as the re-introduced trumpeter swan populations of southern Ontario, Minnesota and Iowa have thrived, residents of the eastern Great Plains and the Missouri–Mississippi valleys can now increasingly observe flocks of several hundred wintering trumpeter swans at several of Iowa's and Missouri's federal wildlife refuges. There they add to the massive migratory flocks of ducks and geese that we residents of the Great Plains have long been privileged to see every autumn and spring.

Seeing immaculate white trumpeter swans flying above the winter-brown loess hills of the Missouri Valley, their elongated bodies and wings outlined gracefully against cerulean skies, always emotionally transports me back to my teen-age North Dakota youth. Then I would wade knee-deep in just-thawed prairie marshes, and stand speechless with awe as wild choruses of whistling swan and snow goose flocks flew endlessly overhead, headed for some unknown breeding places that I knew I must someday try to experience for myself.

With these indelible memories in mind, I recently decided I must revisit in words my beloved waterfowl one last time, as a kind of "swan song," in grateful thanks for the life-long esthetic pleasures, and countless lessons in survival, that these wonderful waterfowl have so freely bestowed on me.

Most of the present text is a variably updated mixture of several of my waterfowl monographs, namely *Handbook of Waterfowl Behavior*, *Ducks Geese and Swans of the World*, and *Waterfowl of North America*. In the several decades since these books were published, several important monographs on waterfowl have appeared. These include Frank Todd's *Natural History of the Waterfowl*, Janet Kear's *Ducks, Geese and Swans* and Guy Baldassarre's masterly revision of F. H. Kortright's classic *Ducks, Geese and Swans of North America*, as well as a few recent books dealing specifically with swans. I have included an extended bibliography to provide a minimum mention of these newer literature sources, although time has not permitted me to more fully update the text. I leave that to some younger person who perhaps is only now beginning to understand and be enamored by the wonders and beauties of ducks, geese and swans and will attempt to help solve their many remaining secrets.

Paul A. Johnsgard
February, 2016

I.

Introduction to the Swans of the World

Because of their immaculate white plumage and their strong pair and family bonds, swans have also long served as icons of beauty, devotion, and longevity in the myths and folklore of many cultures. One's personal interests in and perceptions of wild swans are often formed in childhood, by reading such classics as Hans Christian Anderson's stories of *The Ugly Duckling* and *The Wild Swans,* E.B. White's *The Trumpet of the Swan,* or perhaps upon first seeing a performance of Tchaikovsky's *Swan Lake* ballet.

Many of the most admired human traits, such as permanent pair-bonding, extended bi-parental care, and prolonged family cohesion, are biological facts in swans, although the romantically compelling idea of a dying swan uttering a final "swan song," is sadly only folklore. Yet, a famous American biologist, D. G. Elliott, reported in 1898 that once, after he had shot and wounded a swan in flight, it began a long glide while uttering a series of "plaintive and musical" notes that "sounded at times like the soft running of the notes of an octave" as it gradually drifted downward. Nowadays such unusual behavior would probably be interpreted as being an instinctive distress call, but might have provided an early factual basis for this commonly used expression in describing an individual's final effort.

Most Americans are probably personally familiar with the regal-looking mute swan of Europe, which has long been imported to American parks and zoos. When captive mute swans escaped from Long Island estates during hurricanes in the late 1930's many became feral, and their offspring have since expanded over much of the Atlantic Coast, from New Hampshire to the Carolinas. Introduced mute swans have also spread out from the eastern shoreline of Lake Michigan, occupying much of the Great Lakes region from Wisconsin to New York, and are now classified as an invasive species in several of these states. Yet mute swans are undeniably beautiful, and are likely to be the only species of swan encountered by most Americans, except for those persons living along major swan migratory routes or wintering areas.

Swans are part of the large and worldwide waterfowl family of ducks, geese and swans (Anatidae), which collectively total about 150 species. Most of the waterfowl are important game birds (about 15 million are annually shot in North America), although for aesthetic reasons the swans have traditionally been protected from hunting in nearly all civilized societies. The United States is a notable exception to this generality; here the trumpeter swan had been shot to near-extirpation by early in the 20th century, and several thousand tundra swans still annually fall victims to federally sanctioned "sport" hunting.

All swans, as well as geese, their closest relatives, are moderately to extremely large waterfowl—the mute swan is probably the heaviest of all flying vertebrates. Swans and geese are mostly associated with temperate to arctic climates. They also all exhibit plumage patterns that lack brilliant plumage coloration and are alike in both sexes. The majority of the world's seven species of swans are found in the cooler parts of the northern hemisphere, the exceptions being three southern hemisphere temperate-zone swans.

Most species of swans are seasonally migratory, exhibit prolonged periods of sexual immaturity (typically two or three years), and have strong monogamous pair and family bonds that often persist for several years, if not for their entire adult lifetimes. Swan family bonds are also strong, and may likewise last until their offspring establish pair bonds.

All swans are primarily vegetarians, obtaining much of their food from subsurface aquatic vegetation. The patterns of their downy young tend to be pale and simple, without strong patterning, and in most swans their adult plumages are also fairly simple, with whites and blacks predominating. The white plumage of most adult swans might be related to visual advertisement signals associated with their usual high degree of territoriality.

All pair-bonded swans perform various pair-specific behaviors (see figs. 1-4), These almost always include a mutual "triumph ceremony." This important pair-bonding display is typically performed after an aggressive encounter with another individual, and has the outward appearance of representing a kind of victory celebration. All swans also perform generally highly stereotyped pre-copulatory and post-copulatory postural signals, the latter often accompanied by calling and wing-raising or wing-flapping. Such behaviors that perhaps help in strengthening pair-bonding tend to be much more similar among closely related groups than are the more highly species-specific pair-forming behaviors of such groups that might be more important in preventing interspecific pairing (Johnsgard, 1965).

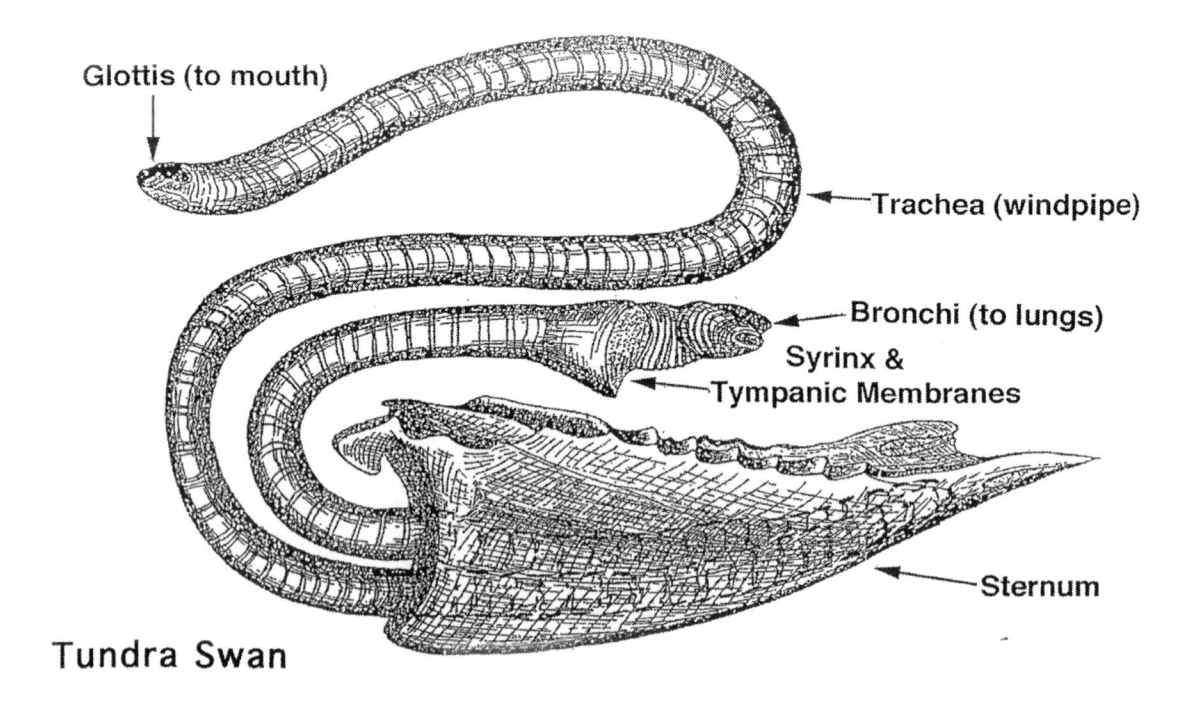

Tracheal anatomy of tundra swan

The most highly territorial swans are also the most strongly vocal species. Three northern swans, the trumpeter, whooper and tundra, have the loudest and most penetrating voices. Their voices are amplified and resonated by virtue of having extremely elongated windpipes (tracheae) that uniquely convolute within the keel of the breastbone (sternum). The resulting increased length and air-chamber volume provide enhanced resonating and harmonic potentials for vocalizing.

In swans, as well as in geese, the voice-producing structures are a pair of highly vibratory ("tympanic") membranes at the junction of the trachea and the two bronchi that connect directly with the lungs (Johnsgard, 1972). Through varied muscular tensions on these paired membranes, complex sounds are generated by expelled air, much like sounds produced by double-reed musical instruments. Swan vocalizations are very similar in both sexes, differing mainly in pitch and harmonic characteristics. In contrast, among typical ducks and some transitional groups such as shelducks there are sex-based structural and acoustical differences evident in adults. Their vocal differences are probably important in facilitating easy sex recognition, and in promoting rapid conspecific pair-bonds, which are often renewed annually (Johnsgard, 1972).

Adults of nearly swan species produce a wide variety of vocalizations, including defensive hisses, parental contact notes, pre-flight and in-flight calls, as well as greeting and triumph ceremony calls within pairs (Cramp and Simmons, 1997; Limpert and Earnst, 1994; Mitchell and Eichholz, 1994). In most swan species such pair-specific sounds probably provide enough inter-individual acoustic specificity as to allow for individual recognition among mated pairs and families, and perhaps even larger social groups. At the other extreme, the mute swan has highly limited vocal abilities, which in adults are mostly confined to soft hisses, grunts, and groans. Its territories are correspondingly relatively small, and in some protected places mute swans even form rather dense nesting colonies ("swaneries") of up to 100 birds.

The biological meanings and significance of most waterfowl vocalizations and other social behaviors are still largely mysteries, as are many other aspects of swan biology, especially in the case of the little-studied swans of South America. Perhaps this brief summary of what is currently known might encourage others to add to our present knowledge.

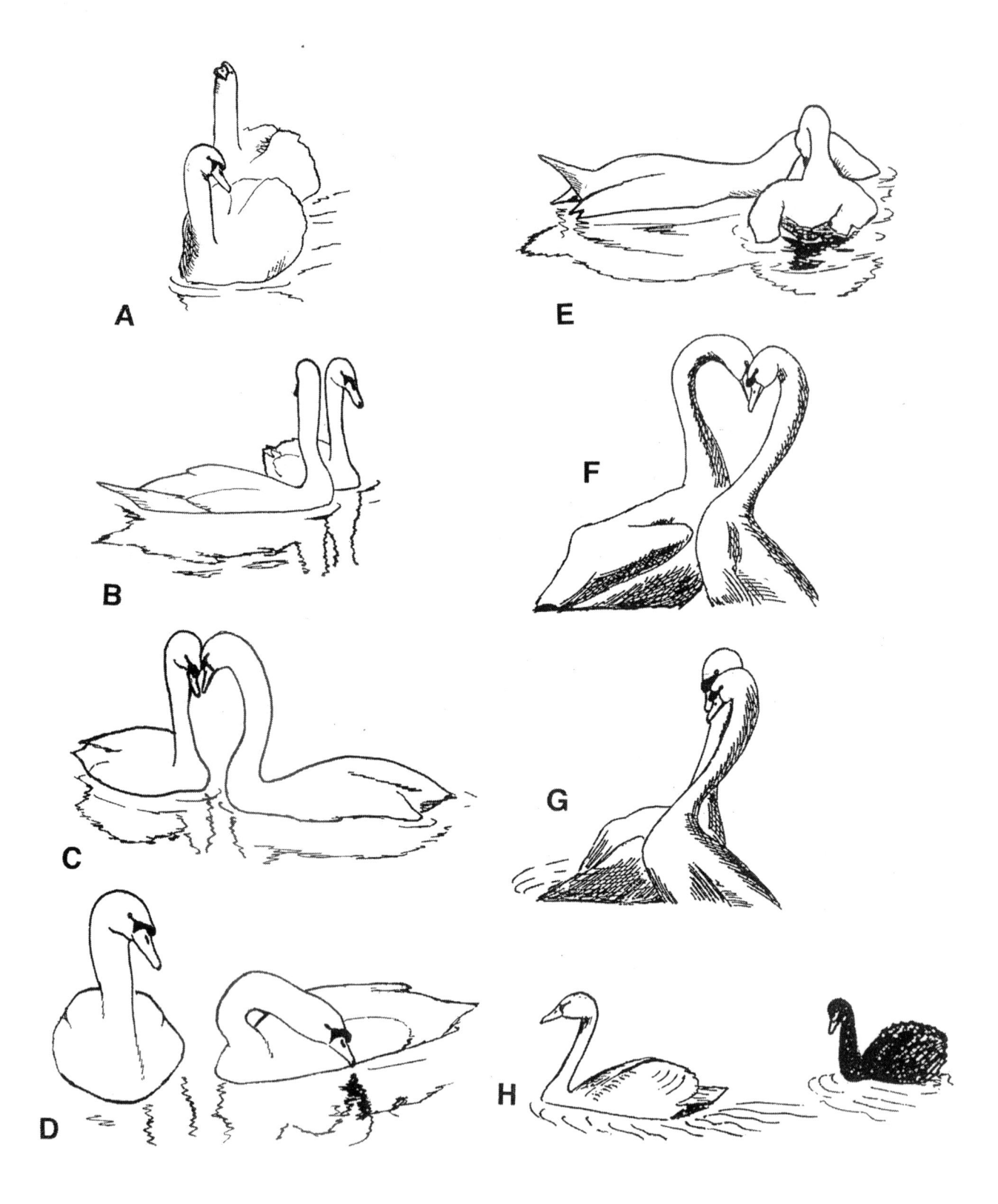

Fig. 1. Behavior of the mute (A–) and black (H) swans: A. Adults in threat posture, rear bird chin-lifting. B–C. Pair-head-turning. D–E. Precopulatory head-dipping. F–G., Postcopulatory display. H. Juvenile mute swan retreating from threatening black swan. (After Johnsgard, 1965)

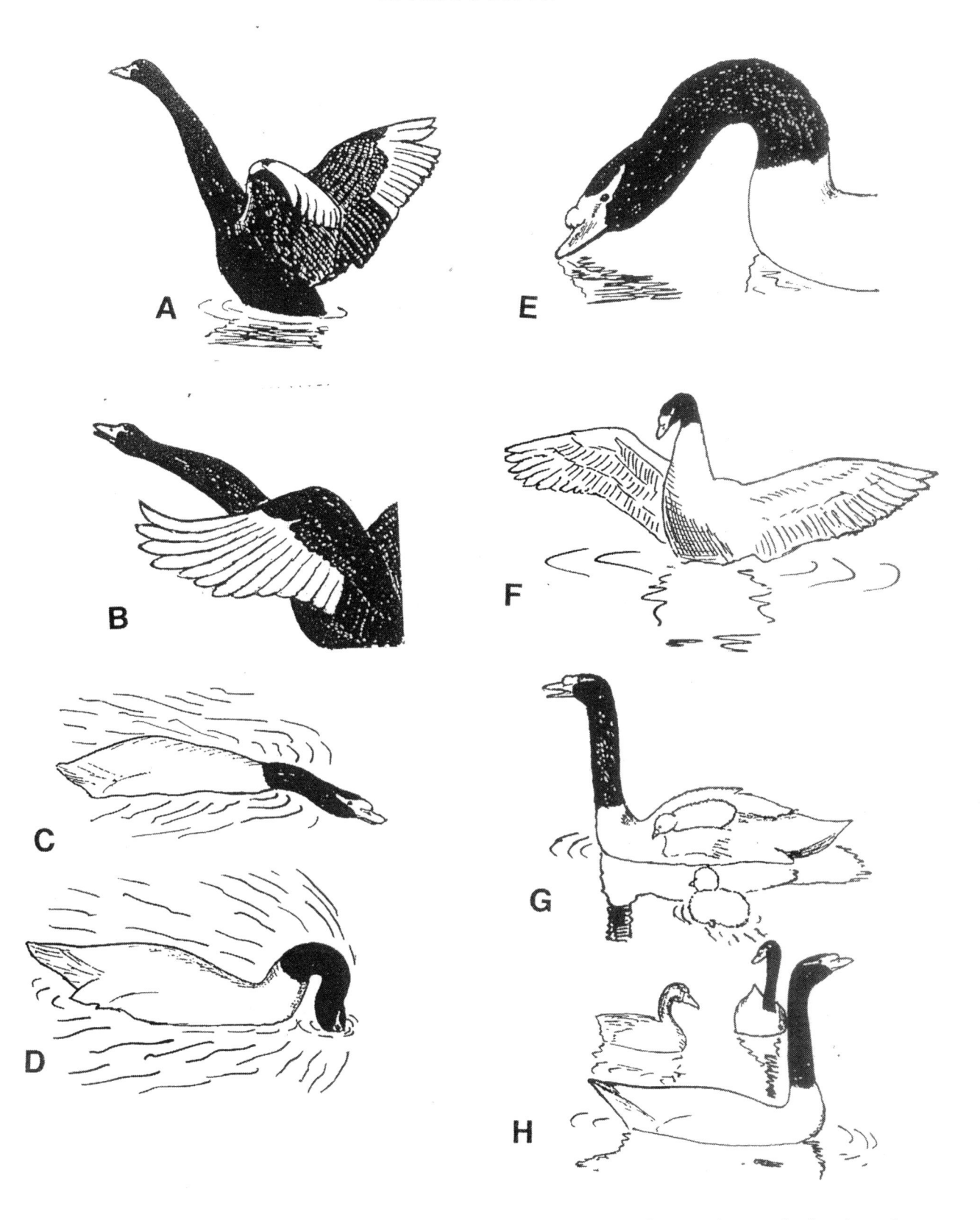

Fig. 2. Behavior of the black (A–B) and black-necked (C–H) swans: A–B. Wing-flapping threat and calling by male. C–E. Male attack (C) and (D–E) final threat posture. F. Male threat with outstretched wings after wing-flapping. G. Female carrying downy cygnet. H. Chin-lifting threat by male, defending female and juvenile. (After Johnsgard, 1965)

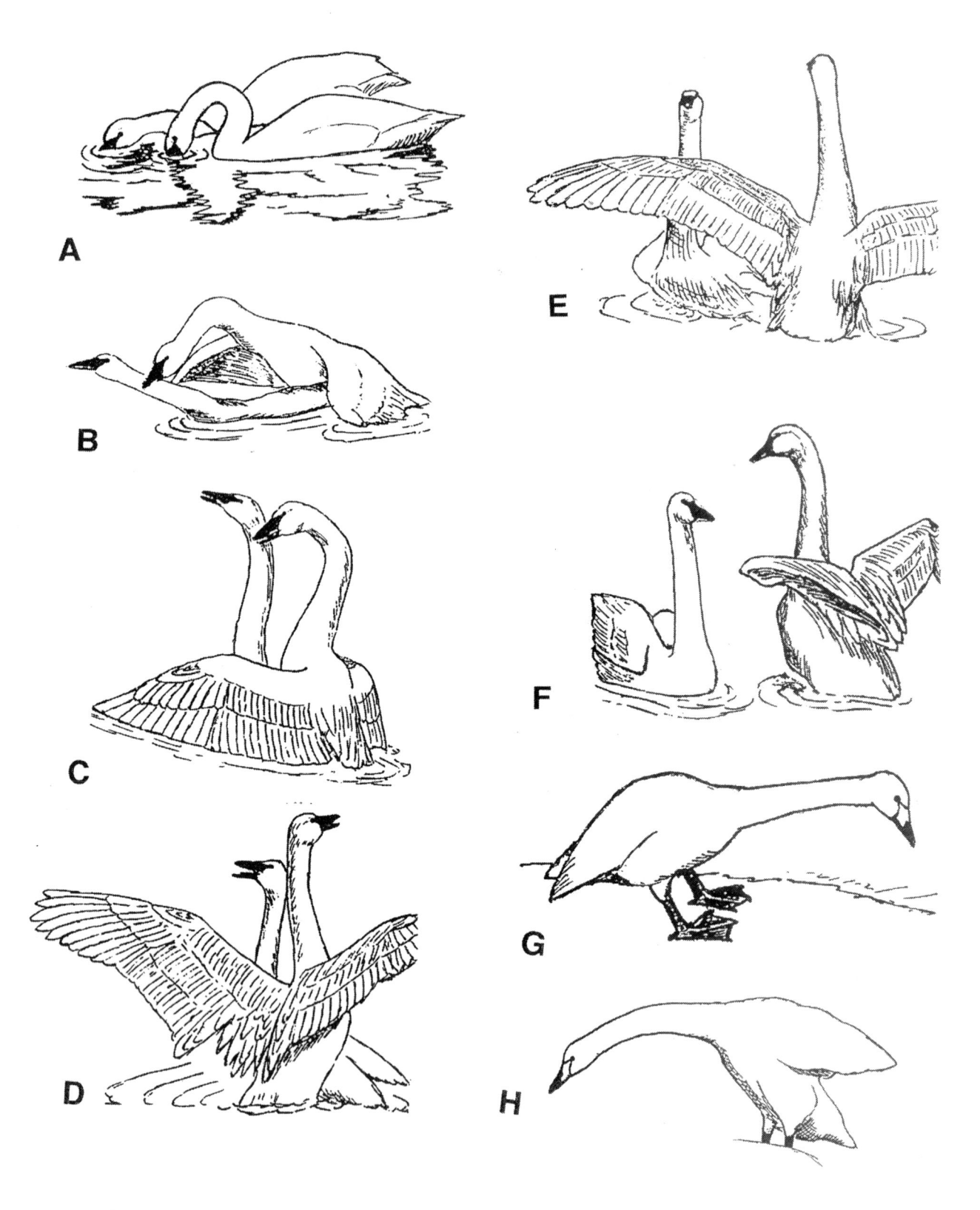

Fig. 3 Behavior of the trumpeter (A–F) and whooper (G–H swans: A–B. Triumph ceremony, male at right. C–F. Post-copulatory display, male has wings extended. G. Male threatening while calling. H. Male performing general shake during threat display. (After Johnsgard, 1965)

Fig 4. Behavior of the whooper (A–B), whistling (C–D), Bewick's (E–F) and coscoroba (G–H) swans: A–B. Triumph ceremony, male at right. C–D. Male threat-calling while wing-waving (C) and while holding wings outstretched (D). E. Triumph ceremony by pair and juvenile. F. Pre-flight neck-bending by family group. G. Threat-swimming by adult male. H. Postcopulatory display, male at left. (After Johnsgard, 1965)

Fig. 5. Mute swan, male in flight

II.

Species Accounts

Mute Swan *Cygnus olor* (Gmelin) 1789

Other vernacular names. White swan, Polish swan; Hockerschwan (German); cygne muet (French); cisne mudo (Spanish)

Subspecies and range. No subspecies recognized. A variant plumage called the "Polish swan" is a mutation-based color morph having a white rather than grayish downy plumage, and pinkish legs and feet that persist from the downy stage into adulthood.

Breeds under native wild conditions in southern Sweden, Denmark, northern Germany, Poland, and locally in Russia and Siberia; also in Asia Minor and Iran east through Afghanistan to Inner Mongolia; in winter to northern Africa, the Black Sea, northwestern India, and Korea. Breeds locally as feral or semi-feral flocks in Great Britain, France, Holland, and central Europe. Introduced and locally established in eastern North America, initially on Long Island and in northern Michigan, but gradually expanding west locally into the Midwestern states (Wisconsin and Illinois), and along the Atlantic Coast from Massachusetts to Georgia, and locally in Florida. Some winter movements southward occur. Also introduced and locally established in South Africa, Australia, and New Zealand.

Measurements and weights. (Mostly from Scott and the Wildfowl Trust, 1972.) Folded wing: males, 589–622 mm; females, 540–96 mm. Culmen: males, 76–85 mm; females, 74–80 mm.

Measurements (after Delacour, 1954): Folded wing: Both sexes 560–625 mm. Frith (1967) reports males as 560–622 and females as 535–570 mm. Culmen: Males (from knob) 70-75 mm. Frith reports males as 70–85 and females as 73–90 mm.

Weights: males, 8.4–15.0 kg (ave. 12.2 kg); females, 6.6–12.0 kg (ave. 8.9 kg). Eggs: ave. 115 x 75 mm, greenish blue, 340 g. Bauer and Glutz (1968) summarized available data. Males seldom weigh over 13.5 kilograms (29.7 pounds), and females should not weigh much over ten kilograms (22 pounds). However, four old birds weighed between September and December averaged 16.225 kilograms (35.78 pounds), with a maximum of 22.4 kilograms (49.39 pounds). Scott & the Wildfowl Trust (1972) presented weight data indicating that although male mute swans average slightly heavier than male trumpeters (12.2 vs. 11.9 kilograms), female mutes average slightly lighter than female trumpeters (8.9 vs. 9.4 kilograms).

Identification and field marks. Length 50–61" (125–155 cm) . *Adults* are entirely white in all post-juvenile plumages. The bill is orange, with black around the nostrils, the nail, and the edges of

Map 1. Breeding (hatching), and wintering (stippling) distributions of the mute swan, excluding introduced populations (from Johnsgard, 1978).

the mandible. The feet are black, except in the uncommon "Polish" color phase, which has fleshy-gray feet. *Females* are smaller (see measurements above) and have a less fully developed knob over the bill. *Juveniles* exhibit a variable number of brownish feathers, which diminish with age (except in the Polish swan variant, which has a white juvenile plumage), and the fleshy knob over the bill remains small through the second year of life. Mute swans are the only white swans that have generally reddish to orange bills adorned with an enlarged black knob at the base (lacking in immatures), outer primaries that are pointed toward their tips, and a somewhat pointed rather than rounded tail shape. The trachea, unlike those of native North American swans, does not enter the sternum.

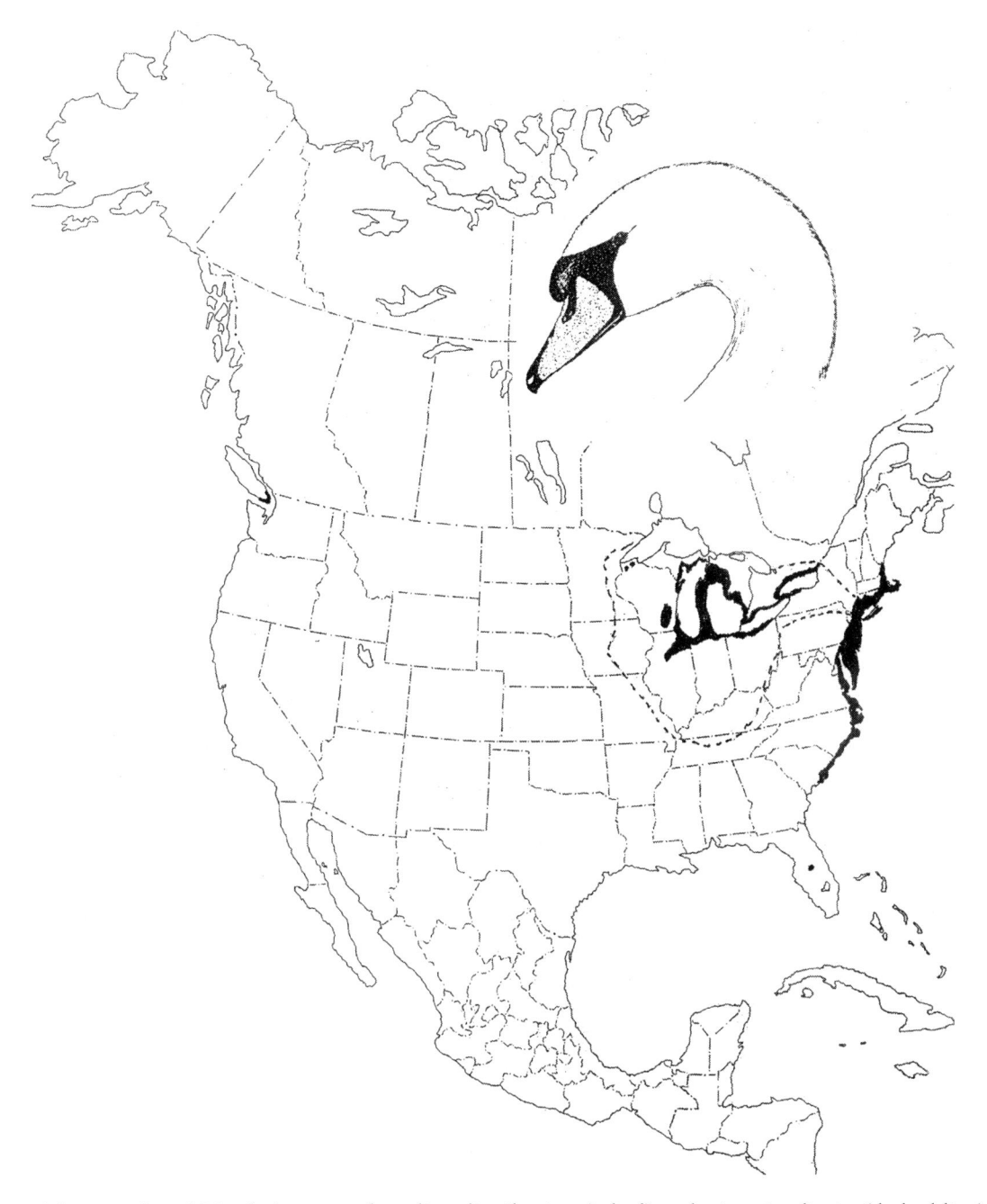

Map 2. The introduced North American breeding distribution (inked) and wintering limits (dashed line) of the mute swan. Small extralimital populations also occur in Florida and along the east coast of Vancouver Island, British Columbia.

Males are considerably heavier and larger than females, and individuals in excess of 10 kilograms are most probably males. Males also have larger black knobs at the base of the bill and most often assume the familiar threatening posture. For immature birds, internal examination is required to determine sex. Any bird still possessing feathered lores or some brownish feathers of the juvenal plumage is less than a year old. Second-year birds may have smaller knobs and less brilliant bill coloration than is typical of older birds.."

Adult male mute swan, mild threat posture.

In the Field: This large swan is usually seen in city parks, but may occasionally be seen as a feral bird under natural conditions, especially in the eastern states and provinces. The neck of the mute swan is seemingly thicker than those of the trumpeter and whistling swans, and while swimming the bird holds it gracefully curved more often than straight. Further, the wings and scapulars are raised when the birds are disturbed, rather than being compressed against the body. The orange bill and its black knob are visible at some distance. In flight, the wings produce a loud "singing" noise that is much more evident than in the native North American swans, and, additionally, mute swans rarely if ever call when in flight, as is so characteristic of the native species. A snorting threat is sometimes uttered by male mute swans, which is their apparent vocal limit.

NATURAL HISTORY

Habitat and foods. The habitats of this species are diverse and numerous, as a reflection of its long association with man and at least partial domestication. Probably, however, marshes, slowly flowing rivers, and lake edges were its native habitats. Clean, weed-filled streams are preferred over larger, polluted rivers. Brackish or salt-water habitats are often used during non- breeding periods.

A variety of studies of foods taken in Europe and North America indicate that the leafy parts of a number of fresh-water or salt-water plants, such as pondweeds, muskgrass, eel-grass *(Zostera),* and green algae *(Enteromorpha),* are major sources of food where they are available. Also where it is available, waste grain is often consumed, and some grazing on terrestrial grasses near water is prevalent in many areas (Scott and the Wildfowl Trust, 1972). Mute swans can reach underwater foods up to several feet below the surface by upending, but in common with other swans do not dive when foraging. A small amount of animal foods, including amphibians, worms, mollusks, and insects, may be taken when available, but these are minor parts of their diet.

Willey (1968) estimated that adults eat an average of 8.4 pounds of vegetation per day. In general, the birds feed on subsurface plants they can reach when swimming or by tipping up in the manner of dabbling ducks. In England these foods include algae *(Chara, Enteromorpha, Ulva, Nitella),* pondweeds *(Zostera, Potamogeton, Ruppia)*, grasses, and other herbaceous plants (Gilham, 1956). Some terrestrial vegetation is also consumed, and sometimes small aquatic animals, including fish and amphibians, have been reported in the diet

In North America mute swans occupy a sedentary breeding and wintering range with habitats that are a directly related to human presence. There seems to be no historical account of the spread of the species in the Hudson Valley and on Long Island after it was originally released as a park bird. Being properly considered an exotic, the species was not included in official bird lists until the 1930s, when the fourth (1931) edition of the *A.O.U. Check-list* noted that it had become established on the lower Hudson Valley and the south shore of Long Island, sometimes straying to the coast of New Jersey.

East Coast hurricanes, such as one that occurred in 1939, caused additional dispersal of birds previously confined to wealthy estates on Long Island and in Rhode Island. By 1949 the species had spread through much of Long Island and had also become well established in Rhode Island. By the late 1950s it was nesting along the entire shore of Rhode Island, and breeding had been reported in the District of Columbia.

A secondary population center was independently developing on upper Lake Michigan, around Grand Traverse Bay and Lake Charlevoix (Edwards, 1966). Early counts of this population were reported by Banko (1960), who noted an increase from two birds in 1948 or 1949 to 41 by 1956. Apparently initiated by a release of two birds in 1918, the flock consisted of at least six hundred by 1973, when efforts began to transplant and establish new flocks in Illinois, Texas, Ohio, Arkansas, Oklahoma, and New Mexico.

In more recent years, pioneering birds have occupied new localities for breeding. These include nestings in Massachusetts, Delaware, New Hampshire, and Connecticut, plus isolated breeding records in South Dakota, Saskatchewan and elsewhere in southeastern Canada and the eastern U.S.A.

The annual Christmas counts of the Audubon Society provide a rough index to the population growth of mute swans in North America. During the years 1949 through 1969, the numbers of such

Adult male mute swan, attack posture.

Adult male mute swan, neck-preening posture.

counts approximately doubled from 403 to 876, while the total number of mute swans counted increased from 374 to 1,644. The average total count for the ten-year period 1950–1959 was 504 birds, with an average of fewer than twenty stations reporting the species, while during the period 1960-1969 the average total count was 1,434 birds, with an average of thirty-four stations reporting mute swans. By comparison, recent surveys suggest that the North American mute swan population was in the vicinity of 22,000–25,000 birds by about 2014. (Baldassarre, 2014).

Social behavior. Flocking occurs among non-breeders and unsuccessful breeders during the midsummer molting period, and later in the fall these flocks are increased by the addition of family groups forced out of their territories by cold weather. Atkinson-Willes (1963) indicated eleven locations (mostly coastal) where accumulations of more than 250 swans have regularly been reported in Great Britain. The largest flocks are generally found on a 1,240-acre reservoir at Abberton, a summer molting area attracting up to nearly 500 birds maximally, and along the Essex coast at Mistley, where 800 to 900 birds are attracted to waste corn from a mill. Except where local foraging conditions favor large concentrations, as on some coastal islands, mute swans are not notably social. Nevertheless, they may flock in groups of a thousand or more in summer molting areas, and similar flocking may also occur in the winter. In many parts of its range the mute swan is essentially sedentary, and in England, for example, studies of banded birds have revealed that most movements are less than 30 miles, and usually follow watercourses (Ogilvie, 1967).

Probably the most extensive migrations occur among the breeding birds of Siberia and Mongolia, from Lake Baikal east. It is presumably those birds that winter along the Pacific coast, from Korea south to the Yangtze Kiang, suggesting a migration of up to about 1,000 miles. Pairs and family units migrate together and remain together until about the end of the year, when at least in England the breeding adults begin to exhibit territorially. At that time the young birds remain in the winter flocks of non-breeding birds, and they may remain together through the following summer and winter (Scott and the Wildfowl Trust, 1972).

Minton (1968) has studied population densities in England and reported a density of one pair (about 30 percent non-breeders) per 5.5 square miles on his study area of 550 square miles. He noted that this represented about one breeding pair per eight square miles, com- pared with earlier estimates of one pair per 16 square miles reported for England and Wales as a whole. The highest reported county densities were one pair per three square miles for Middlesex and one per seven square miles in Dorset. Atkinson-Willes (1963) reported that the famous mute swan colony at Abbotsbury in Dorset averaged 66 pairs of breeding swans (range 39-104) in the years 1947–1956, and had an average total population of about 700 birds. A tradition of protection and abundant food in the form of *Zostera* and *Ruppia* account for this concentration of birds. Comparable figures are not available for North America, but the highest Christmas counts have usually occurred in central Suffolk County, where the total number of birds seen in a 15-mile-diameter area (176 square miles) has averaged 452 for the 1960-1969 period, or 2.6 per square mile. If Minton's estimate that 30 to 40 percent of the population represents breeding birds, this would represent a breeding density of nearly one pair per square mile, assuming no spring dispersal. Willey (1968) estimated that 24.5–54.3 percent of the Rhode Island population represented potential breeders. Thus it would seem that, at least locally, mute swan breeding populations in North America may be as high as or higher than in Great Britain.

In Europe the mute swan is a species that nests largely in populated areas that support few other breeding waterfowl, and there is probably little competition with other species. Dementiev and Gladkov (1967) reported it tolerant toward other birds and sometimes occurring with nesting graylag geese (*Anser anser*). However, Willey (1968) stated that nesting birds sometimes kill other swans that intrude into their nesting areas. He also considered mute swans to be a substantial threat to humans, particularly children. Stone and Masters (1971) reported that six captive mute swans killed six adult geese and two adult ducks, as well as 40 ducklings and goslings, during a 20-month period.

Mute swans are highly sedentary birds in Great Britain. Atkinson-Willes (1963) reported that only a small number of banded mute swans had been proven to move more than a hundred miles, and only two had been known to cross the English Channel. Later, Harrison and Ogilvie (1968) noted that ten of 2,700 band recoveries exhibited overseas movement from Great Britain, with recoveries from Holland, the Baltic coasts of East and West Germany, Sweden, and France. Many of these recoveries were related to severe winter conditions that forced birds to move from the continent to Britain.

According to Minton (1968), most movements of mute swans occur before their mating and acquisition of a territory, after which they become quite sedentary. Most pairs return to their territory year after year, with only two percent of the surviving paired population that Minton studied moving their territories more than five miles. Non-breeding pairs and unsuccessful breeders frequently move to the nearest flock for molting in midsummer, while unsuccessful breeders molt on their territories and move into flocks during fall. Among paired birds, movements are usually less than ten miles, and only about five percent of the 450 pairs studied moved farther than this. However, unsuccessful breeders are more likely to move greater distances than successful ones.

Flocking typically occurs among non-breeders and unsuccessful breeders during the midsummer molting period, and later in the fall these flocks are increased by the addition of family groups forced out of their territories by cold weather. Atkinson-Wills (1963) noted eleven locations (mostly coastal) where accumulations of more than 250 swans had regularly been reported in Great Britain. The largest flocks were generally found on a 1,240-acre reservoir at Abberton, a summer molting area attracting up to nearly 500 birds maximally, and along the Essex coast at Mistley, where 800 to 900 birds were attracted to waste corn from a mill.

Reproductive biology. Pair-forming behavior occurs in the fall and winter, usually during the season prior to the bird's initial breeding. Among mute swans, initial breeding is most frequent in the third year, with some birds (mostly females) breeding when two years old and some not until four or older. Pair formation and pair-bonding in this as in all typical swans occurs by mutual greeting ceremonies such as head-turning, and bonds are firmly established by triumph ceremonies between members of a pair. This occurs after the male has threatened or attacked an enemy" and returned to his mate or prospective mate with ruffled neck feathers and raised wings, calling while chin lifting. Precopulatory behavior may occur frequently among paired birds in winter flocks, and consists of mutual head dipping and preening or rubbing movements along the flanks and back. After treading, both birds rise in the water breast to breast, calling and extending their necks and bills almost vertically for a brief moment, then subside in the water (Johnsgard, 1965a). Once formed, pair bonds are very strong, and so long as both members of a pair remain alive there is a low rate of mate changing. One study in England indicated that there was a three percent "divorce" rate among unsuccessful breeders or non-breeders (Minton, 1968).

The earliest known age of reproductive maturity in North America has been reported as two (Johnston, 1935) or three (Willey, 1968) years, but studies in England indicate considerable variation may occur. Perrins and Reynolds (1967) indicated that three years of age is the most common time of initial breeding for females, but a few birds may breed at two and some may not breed until they are six years old. Initial breeding by males occurred between three to seven years of age. Minton (1968) found that of forty-three mute swans, half initially nested and raised young at the age of three, while an additional third did so the following year, with a slight tendency for females to mature earlier than males. Three birds did not breed until they were at least six years old.

Minton (1968) reported on the initial pairing behavior of 125 mute swans of known age. Nearly half of these were two-year-olds, another 30 percent were three-year-olds, and a few (one male, four females) took mates when only a year old. Most birds were paired for at least a year before they actually attempted to nest, with only two of 60 birds that were no more than two years old actually nesting that year. Birds tended to pair with others of about their own age, with a slight tendency for the males to be older than the females. Further, in 74 percent of the initial pairings neither partner had ever been paired before. Birds pairing for the first time with a previously paired bird were generally replacements for dead mates.

The strong pair bond of all swans is well known and has been well documented by mute swans.. Minton (1968) reported that "divorce" (the changing of partners when both are still alive) among the paired population had an incidence of about four percent for nonbreeding pairs and one percent for breeding pairs. In cases where both birds survived to following years, 82 percent of the successful breeders and 78 percent of unsuccessful breeders remained paired. Of seventy-one pairings first studied in 1961, six were still intact in 1966. During the six-year study, eleven males and nine females were known to have had at least three different mates, but in several cases (twelve males and two females) birds that had apparently lost their mates remained on their nesting territory the following year. In some cases there was a gap of two or three years before re-pairing, while in others the birds apparently gave up pairing permanently.

Copulatory displays have been described by various persons, such as Boase (1959), Johnsgard (1965), and others. Precopulatory displays involve mutual bill-dipping and preening movements, with the neck feathers ruffled. Following treading, both birds rise in the water breast-to-breast, with their necks and heads extended vertically but with their wings closed; then they gradually arch their necks and settle back on the water.

The nesting period of mute swans occurs in spring, which is generally March through June in its northern hemisphere range, and from September through January in New Zealand and Australia. Mute swans are highly territorial, although the size of the territory varies with breeding density. In England, nests are most often placed in or near standing water, less often beside running water, and least often in coastal situations. Nests are sometimes grouped in colonies, with one colony in Dorset having had as many as 500 nests (Scott and Boyd, 1957). One study in Staffordshire, England, indicated a density of a pair per 2,000 hectares, but with territories limited to a few meters of streamside locations. In a few areas of dense colonies, the nests may be located only a few meters apart.

Established breeders tend to use previous nest sites. Willey (1968) estimated the average size of twelve nesting territories as 4.4 acres (range 0.5–11.8) in Rhode Island. Minton (1968) noted that both breeding and nonbreeding pairs were more prevalent on small (ten acres or less) water areas than on

Adult mute swan pair, male standing beside sleeping female.

larger ones, but considering availability, larger water areas were slightly favored. Likewise, streams were favored over canals or rivers (over 20 feet wide), especially by breeding pairs. Clean waters with aquatic vegtation were also preferred over more polluted waters.

Most studies indicate that about six eggs constitute an average clutch size for mute swans; Perrins and Reynolds (1967) reported such an average for 92 nests. Studies summarized by Bauer and Glutz (1968) also indicate averages of between 5.8 and 6.2 eggs. Clutch sizes of up to 11 eggs laid by one female are known, but renesting attempts appear to average smaller clutches, of about four eggs (Perrins and Reynolds, 1967).

After the establishment of a breeding territory, nests are constructed on land or shallow water. The nests are usually about a meter in diameter and 0.6 to 0.8 meter in height and are constructed in the form of a large mound of vegetation consisting of rushes, reeds, other herbaceous vegetation, and sometimes also sticks. The nest cup is lined with finer materials and also with down and feathers. The female typically does most of the nest construction, but the male also gathers material from nearby, passing it back toward the nest over his shoulder. Down-plucking may begin with the start of egg-laying, the initiation of incubation, or not until the last or penultimate egg is deposited. The female does the incubation, but is closely guarded by the male. The cygnets typically leave the nest on the day after hatching and remain closely attended by both parents, often riding on the backs of one or both parents. The wing molt of both parents normally occurs during the fledging period of the brood (Bauer and Glutz, 1968; Dementiev and Gladkov, 1967, etc.).

Nests are large piles of herbaceous vegetation, built by both sexes, and often built on the previous year's nest, especially if it had been a successful site. Eggs are deposited every other day until a clutch of 4–8 (usually 5–6) eggs is laid. Incubation is normally performed only by the female, but the male may occasionally take over for a time.

The incubation period has generally been estimated as 35 or 36 days, with some estimates of up to 38 days (Bauer and Glutz, 1968). The female incubates, but the male actively protects the nest. Minton (1968) reported a 59 percent nesting success among 352 pairs, and a 52 percent success for 11 renesting attempts, with 80 percent of the nest losses due to human disturbance or destruction. Willey and Halla (1972) reported the loss of 87 eggs and young from a total of 47 nests after severe flooding and cold weather in Rhode Island.

After hatching, both sexes attentively care for the young, frequently allowing them to ride on their backs. The fledging period has been variously reported as four and a half months (Bauer and Glutz, 1968), 18 weeks (Lack, 1968), 18–20 weeks (Scott and Boyd, 1967), and 120–150 days (Kear, in Scott and the Wildfowl Trust, 1972), during which time the adults undergo their postnuptial molt.

Successful breeders remain with their young well past the cygnets' fledging time, usually until severe weather forces the families to retire to winter quarters and to merge with larger groups of swans. Typically, the young of the past year are driven out of the territory by their parents before the latter begin to breed again. Minton (1968) reported two cases in which young remained with their parents until the following summer or until molting, and in neither case did the parents breed during that year. Minton observed two cases of pairing between parents and offspring. One involved the pairing of a female with its yearling son after the male parent had died, while the other involved a female observed paired with a two-and-one-half- year-old son. In neither case did actual nesting occur.

Minton (1968) found that the average brood size (219 broods) at fledging over a six-year period was 3.5 birds, while the total number raised to fledging averaged 2.0 per breeding pair. Perrins and Reynolds (1967) found an average brood size of 3.1 young for 83 broods, with an estimated 2.0 young raised per pair to September, including pairs that did not hatch any young at all. They estimated that the average mortality rate between hatching and fledging was 50 percent, with an additional 23 percent mortality rate for the rest of the year. Willey (1968) estimated a pre-fledging mortality of 56.4 percent in 1968, with the snapping turtle (*Chelydra serpentina*) apparently a primary predator of cygnets. After fledging, the family increases their food intake and fat reserves before leaving the breeding grounds

Perrins and Reynolds (1967) estimated that among immature birds there is a 67 to 75 percent survival (25 to 33 percent mortality) rate, while breeding adults have a survival rate of 82 percent, possibly decreasing after the sixth year of life. There is little difference in the estimated mortality rates of the two sexes. Ogilvie (1967) estimated a higher mortality rate of 40.5 percent for birds banded when under a year old and 38.5 percent for those banded when over a year old, with the possibly greater survival in the third and fourth years of life than during the first two. Overhead wires were found to be a major cause of mortality, with oiling, disease, fighting, cold weather, and shooting also accounting for some mortality. Maximum longevity record in captivity is 21 years (Scott & the Wildfowl Trust, 1972).

Status. This swan has greatly benefited from man's influence, and has either been purposefully introduced or otherwise spread into new breeding areas in historical times. In many areas of Europe where the species was exterminated during the 1800s it is now reestablished and increasing.

Relationships. In spite of its similar geographic range and plumage characteristics, the mute swan is not closely related to the other northern hemisphere white swans, but instead its nearest relative is the black swan, as is indicated by a variety of behavioral characteristics (Johnsgard, 1965a).

Suggested readings. Scott & the Wildfowl Trust, 1972, del Toyo, Elliott & Sargatal, 1992, Todd, 1996. Ciaranca, Allin & Jones, 1997, Kear, 2005, Baldassarre, 2014.

Black Swan *Cygnus atratus* (Latham) 1790

Other vernacular names. None in general English use. Trauerschwan (German); cygne noir (French); cisne negro (Spanish).

Subspecies and range. No subspecies recognized. The native breeding range includes Australia, except for the north and central areas, and Tasmania. Winters over most of its range. Introduced into New Zealand and well established on both islands.

Measurements and weights (mainly from Frith, 1982). Folded wing: males, 43–543 mm; females, 416–99 mm. Culmen: males, 57–79 mm; females, 56–72 mm. Weights: males, 4.6–8.75 kg (ave. 6.27 kg); females, 3.7–7.2 kg (ave. 5.1 kg). Eggs: ave. 115 x 65 mm, pale green, 300 g.

Identification and field marks. Length 45–55" (115–40 cm). This is the only black swan; *adults* are uniformly dark brownish black, with somewhat paler underparts and white flight feathers. All the primaries and the outer secondaries are white, while the inner secondaries are white-tipped. The innermost wing feathers are strongly undulated. The bill is red to orange, with a whitish sub-terminal bar and nail; the iris is reddish (or sometimes whitish), and the feet and legs are black. *Females* are smaller and have a less brightly colored bill and iris. Also, they appear to have shorter neck feathers and a less ruffled back, which may simply be the result of variable feather lifting. *Juveniles* are a mottled grayish brown, with light-tipped feathers and a paler bill coloration.

In the field, the blackish plumage makes this species unmistakable for any other swan. In flight, the white flight feathers contrast strongly with the otherwise blackish coloration. The call of the adults is a rather weak bugling sound that does not carry great distances.

NATURAL HISTORY

Habitat and foods. The preferred habitat of black swans in Australia consists of large, permanent lakes, of either fresh or brackish water. Less often they are found on rivers and billabongs, and occasionally occur along the coast. Water depth is more important than water chemistry; abundant aquatic vegetation must be available to within three feet of the surface to provide a food source. Foraging is done primarily by upending in water, but sometimes the birds graze along the banks or stand in shallow water and dabble. Of the foods eaten by a sample of birds from New South Wales, all were of vegetable materials, except for a few items of animal origin that may have been accidently ingested with the plants. Over a third of the volume of food consisted of cattails, while the other materials that occurred in large quantities were algae, wild celery *(Vallisneria)*, and pondweeds. A variety of grasses and their seeds also were part of the sample. Most of the food materials were of aquatic plants, with shoreline plants contributing only a very small percentage (Frith, 1982). Less is known of the foods consumed in New Zealand, but apparently the birds spend more time grazing in pastures there, and sometimes even strip and eat the leaves of willows.

Social behavior. The black swan is one of the more social of the swans, perhaps in part because of lack of definite territorial behavior, which favors the development of large breeding colonies as well as concentrations in the nonbreeding season. Many lakes in southern Australia regularly support from 5,000–15,000 swans, and on Lake Ellesmere of New Zealand estimates of as many as 60,000–80,000 birds have been made.

Such large flocks are of dynamic composition, but comprise the stable pair and family units of all swans. Further, the flocks are probably more mobile than is generally thought to be the case; banding studies of birds caught while molting in Australia indicate movements of up to several hundred miles in successive years. These studies indicate that most swans are essentially nomadic, moving widely as weather and availability of suit- able habitat dictate.

Pair formation presumably usually occurs during the second winter of life, although this has yet to be verified. As in the mute and other swans, pairs are formed and maintained by various mutual displays, particularly the triumph ceremony. This ceremony takes essentially the same form in the black swan as in the mute, with strong chin lifting, mutual calling, and ruffling of the neck feathers. Precopulatory behavior consists of mutual bathing movements but no preening, and after treading, the birds

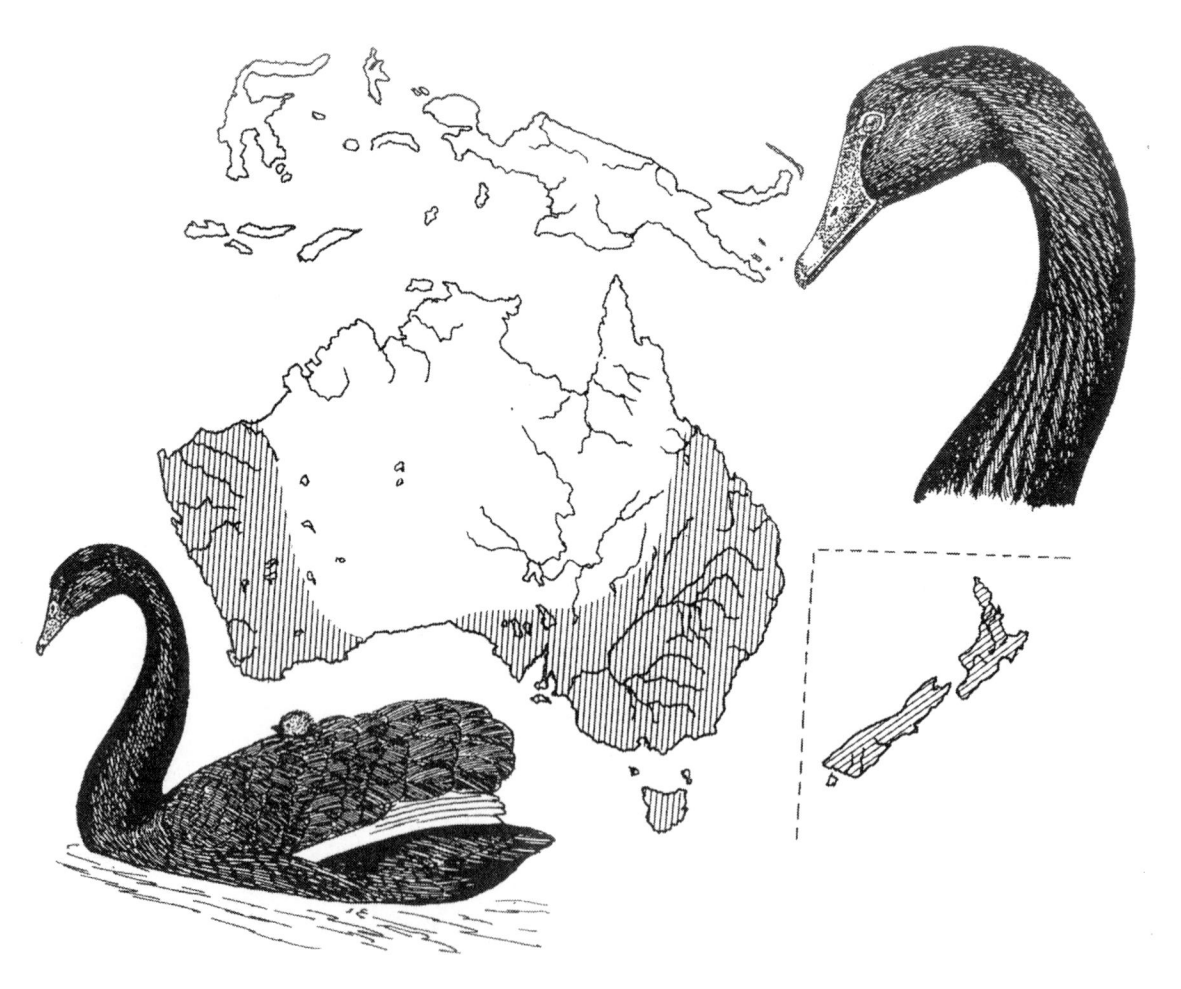

Map 3. Native (vertical hatching) and introduced (New Zealand only) distributions of the black swan (from Johnsgard, 1978).

Fig. 6. Black swan, adults taking flight

do not rise breast to breast, but rather extend their necks and heads vertically, call once, and then swim about in a partial circle before starting to bathe.

Reproductive biology. The timing of breeding in black swans is fairly variable. In its native range of Australia the birds breed in the rainy season of February to May in Queensland, and from June through August in western Australia, during that area's moister winter. In the central and southern areas breeding may be timed according to local rains, or during June and July in more permanent water areas (Frith, 1982). In New Zealand most breeding occurs from August to November, and in some areas nesting also occurs in the austral autumn. Among captive birds in Europe, records of breeding show a spread from March through September, but about 70 percent of the records are for April and May (Petzold, 1964).

Nests are built either on land or in swamps, and also sometimes are placed on stumps, at the bases of trees, or in floating debris. In swamps of cattail they consist of large clumps of these plants up to five feet wide and three or four feet above the water surface. Ground nests are appreciably smaller. Frequently the previous year's site is used; in Australia it has been found that a site may be used as many as four times. In large colonies, the nests may be as close as a meter apart (Guiler, 1966), or practically touching each other. It has been suggested that the slow growth rate of the cygnets, their mostly vegetarian diet, and perhaps their increased protection from predation have facilitated the evolution of colonial nesting in this species (Kear, in Scott and the Wildfowl Trust, 1972).

Black swan, brooding female and cygnets. Note female's erected neck feathers.

Both sexes help build the nest, and a clutch of 4–10 eggs (most often 5) is laid on an alternate-day basis. Clutch sizes vary by date and location, and in Tasmania were found to average highest in years of poor breeding conditions, when only experienced breeders attempted to nest.. The male regularly helps incubate in this species, which is unique among swans. The two sexes shift incubation duties at about 3–4 hour shifts, with the female usually incubating at night. All told, the male averages more incubation time than the female. Under natural conditions the average incubation period is about 40 days, but it ranges from 35–45 days (Frith, 1982).

The young grow fairly slowly, at least as compared with the swans that nest in the arctic, and 140–180 days elapse before the flight feathers are fully grown (Frith, 1982). In New Zealand the chicks fledge in 95–140 days, depending on the availability of food. Brood mergers are frequent when several broods forage within a limited area. Such multi-family groupings may persist until near the end of the fledging period (Kear 2005). The maximum longevity record in captivity is 33 years (Scott & the Wildfowl Trust, 1972).

Black swan pair, in alert swimming posture.

Status. The black swan's status in Australia is very favorable; it is currently protected and has recently extended its breeding range in southern Queensland. Some short hunting seasons have been needed in some years to cope with local crop depredation problems in Victoria and Tasmania. Likewise, in New Zealand, where it was introduced in 1864–68, the species is locally very common, and around Lake Ellesmere intentional disturbance and commercial egg collecting control its numbers. In Australia surveys of wetlands in eastern Australia from 1983–1988 resulted in wildly varied estimates of 13,800–156,000 (Marchant & Higgins, 1990), probably as a reflection of the unpredictable amounts and durations of Australia's annual precipitation. About 60,000 were estimated to be present in New Zealand in 1980, and about 3,000 were also present on the Chatham Islands.(Williams, 1981).

Relationships. The black swan is clearly a close relative of the mute swan, in spite of the differences in adult plumage coloration (Johnsgard, 1965a). Livezey (1996) placed the black and black-necked swan in a subgenus of southern hemisphere swans that he called *Chenopsis*. A possible reason for the evolution of a dark plumage in the black swan may be a reduction in ecological needs for conspicuousness in conjunction with the effective advertisement and defense of a large territory, which seem to be a primary reason for whiteness in the other swan species.

Suggested readings. Guiler, 1966; Frith, 1982; Scott and the Wildfowl Trust, 1972; Marchant and Higgins, 1990; del Toyo, Elliott & Sargatal, 1992; Todd, 1996; Kear, 2005.

Black-necked Swan *Cygnus melancoryphus* (Molina) 1792

Other vernacular names. None in general English use. Schwarzhalsschwan (German); cygne a col noir (French); cisne de cuello negro (Spanish).

Subspecies and range. No subspecies recognized. Breeds in Paraguay, Uruguay, Argentina, Chile, Tierra del Fuego, and the Falkland Islands. Winters as far north as the Tropic of Capricorn, in Paraguay and the three southern provinces of Brazil.

Measurements and weight (mainly from Scott & the Wildfowl Trust, 1972). Folded wing: males, 435-50 mm; females, 400–415 mm. Culmen: males, 82–86 mm; females, 71–73 mm. Weights: males, 4.5–6.7 kg (ave. 5.4 kg); females, 3.5–4.4 kg (ave. 4.0 kg). Eggs: ave. 105 x 65 mm, cream, 247 g.

Identification and field marks. Length 45–55" (115–40 cm). This swan is the only species that is entirely white except for a black head and neck. There is also a large red caruncle behind the bill, which is bluish gray with a paler nail, and the legs and feet are pink. *Females* are noticeably smaller than males, but have well-developed caruncles behind the bill. *Juvenile* birds exhibit a variable amount of brownish gray in their plumage, which is lost by the end of the first year, after which immatures may be distinguished by their smaller or absent caruncles.

In the field, the contrasting white and black body plumage is unmistakable; the coscoroba swan is the only other swan in the native range of this species. The black-necked swan's voice is a relatively weak and wheezy whistle, and does not carry a great distance.

NATURAL HISTORY

Habitat and foods. This is a species of fresh-water and brackish-water marshes, predominantly the former. It is mostly found at lower altitudes, but at times also occurs on upland lakes. Weller (1975) found the birds on large marshes or marshy lake edges in southern Argentina. Near Buenos Aires Weller (1967) saw them in lakes with tule (*Scirpus*) edges or in large pools in marshes. In a Chilean study, the primary foods taken were the leaves of *Egeria densa*, the dominant submerged aquatic plant (Schlatter *et al.*, 1991) On the Falkland Islands the swans favor large waters (over 40 hectares) that may be either brackish estuaries or fresh, but they spend little time in kelp beds (Weller, 1972). They were found where algae *(Nitella)* and mud plantain *(Heranthera)* occurred, presumably feeding on them both. However, very little is known of the specific foods taken by this swan. It is rarely seen ashore, and thus must be almost entirely dependent on aquatic plant materials for its diet. Johnson (1965) reported that aquatic insects and fish spawn may sometimes also be eaten.

Social behavior. This is a relatively gregarious swan, with nonbreeding birds forming flocks of up to several thousand birds seasonally. Weller (1967a) found a flock of 5,000–6,000 birds on an Argentine marsh in midsummer after the young had fledged, but generally the observed flock sizes are much smaller. Migratory movements in South America are no doubt fairly substantial, and the birds obviously

Map 4. Breeding or residential (hatching) and wintering (stippling) distributions of the black-necked swan (from Johnsgard, 1978). Marginal wintering range is shown by a dashed line.

can fly long distances, as attested by an occasional stray reaching the Juan Fernandez Islands, 400 miles off Chile's coast. Even on the Falkland Islands there are some seasonal movements evident (Cawkell & Hamilton, 1961). Probably most of the northward movement from central and southern Argentina occurs in March and April, after the year's young are well fledged. Formation of pairs almost certainly takes place in winter flocks; two-year-old birds often breed in captivity, but it is assumed that under natural conditions the birds probably begin breeding when approaching the end of their second year. Captive-raised birds may breed when only two years old (Kear 2006), although studies of wild birds in Chile indicated that only 7.6–16.2 percent of the population breeds every year, with brood sizes ranging from 1.5–2.8 young produced per pair.

Fig. 7. Black-necked swan, adult landing

Pair formation is achieved by the repeated use of the triumph ceremony; although this is the smallest of the swans it is among the most aggressive, and males of pairs in captivity are almost constantly threatening other birds, then returning to their mates and performing a chin-lifting triumph ceremony. Unlike the mute and black swans, black-necked swans to not raise their folded wings in threat, but extend their neck and head low over the water as they rush toward the opponent with repeated calls. Because of this high level of aggression, it seems unlikely that coloniality is normal in this species, and most observers indicate a low breeding density. Cawkell and Hamilton (1961) reported six to eight pairs on a large lake in the Falkland Islands, and Weller (1967a) found one nest and saw several broods on a 100-hectare lake near Buenos Aires. A cluster of six nests (of which four were empty) were once found within 18 meters of one another on the Falkland Islands. In Chile the average distance between nests along Rio Cruces was 13.6 meters (Schlatter *et al.*, 1991), in an area where few nesting opportunities were present elsewhere.

Reproductive biology. Black-necked swans nest in the southern spring, and were the earliest marsh-nesting species Weller (1967) studied in east-central Argentina. He noted that birds apparently started nesting there in July. Likewise in central Chile breeding begins in July and August, but farther south it may begin at least a month later. On the Falkland Islands nesting occurs from early August to mid-September. In Chile the clutch size varies from 4 to 8 eggs, and the nest is placed in thick reed beds around lake edges or lagoons; preferentially it is placed on small islets (Johnson, 1965). It is usually large and bulky, often built of rushes and partially floating. The male closely guards the nest and stands over it when his mate leaves to forage, but does not normally help incubate.

The incubation period is usually 35 days (34–36) days, with the female leaving the nest during the evening hours every few days for foraging. Dominican gulls *(Larus marinus dominicanus)* are reportedly a

Nesting black-necked swan pair, mutual chin-lifting and calling ceremony.

Black-necked swan family, mutual chin-lifting.

serious egg predator in the Falkland Islands (Cawkell & Hamilton, 1961), and common caracaras (*Polyborus plancus*) are similarly serious predators on both eggs and chicks in Chile (Schlatter *et al.*, 1991). The newly hatched cygnets are closely attended by both parents, and more time is spent carrying the young on the backs by both adults than is true of any of the other swans. Indeed, the male sometimes does the majority of such carrying of young. The strong degree of parental carrying in this species may be associated with the fact that these swans rarely come ashore, and thus terrestrial brooding is impossible (Kear, in Scott and the Wildfowl Trust, 1972).

There is no specific information on the length of time to fledging among wild birds in this species, but the fledging period of captive-raised birds is about 16 weeks (Kear, 2005). The growth rate is obviously very slow, apparently second only to that of the black swan, which requires nearly six months to fledge. Weller (1967a) reported that in central Argentina young had fledged by early January, while hatching occurred in October, suggesting a fledging period of about 100 days. During this fledging period the adults undergo their annual postnuptial molt, and departure from the breeding grounds occurs shortly after the young are fledged. The maximum longevity record in captivity is 20 years (Scott & the Wildfowl Trust, 1972).

Status. In Chile, the black-necked swan has recently reoccupied a part of its range that had previously been eliminated as a result of local persecution, and in Argentina the swan is locally common in many marshy areas. Destruction of marshes, either by their removal with drainage canals or by use of the marshes as a runoff repository for excessive rainfall on adjacent agricultural lands, rather than specific hunting or other persecution of the species is the biggest threat to its long-term survival. Periodic droughts are also considered a significant problem. Few detailed population estimates are available, but the species' total numbers may have exceeded 100,000 in the early 2000s, with 20,000 in Chile, 50,000 in Argentina, at least 20,000 in Uruguay, 2,000–3,000 in southern Brazil. and 750–1,500 in the Falkland Islands (Kear, 2006).

Relationships. Although it is relatively isolated from the others of this group, the black-necked swan is fairly clearly a member of the mute swan and black swan evolutionary complex. It differs in a number of behavioral ways from these two species, however, and except for the coscoroba is perhaps the most isolated in an evolutionary sense of all of the swans.

Suggested readings. Johnson, 1965; Weller, 1967; Scott and the Wildfowl Trust; 1972; Schlatter *et al.*, 1991; del Toyo, Elliott & Sargatal, 1992; Todd, 1996; Kear, 2005.

Trumpeter Swan *Cygnus buccinator* (Richardson) 1758

Other vernacular names: Wild Swan.

Range: Native breeding populations currently exist in southern Alaska, British Columbia, western Alberta, eastern Idaho, southwestern Montana, and Wyoming. Reintroduced and breeding at various national wildlife refuges in Oregon, Washington, Nevada, South Dakota, and eastward from Ontario, Minnesota and Iowa to Michigan, Ohio and New York, and increasingly occurring to New England. Some movements occur in winter south to California, New Mexico and Texas, but most populations are not strongly migratory. Considered by Delacour (1954) only subspecifically distinct from *C. cygnus*, but recognized by the American Ornithologists' Union (1998) as a separate species.

Measurements (after Banko, 1960). Folded wing: Adult male 545–680 mm. (average 618.6), adult female 604–636 mm. (average 623.3). Culmen: Adult male 104–119.5 mm. (average 112.5), adult female 101.5–112-5 mm. (average 107). Weights: Nelson and Martin (1953) indicated an average weight of seven males as 27.9 pounds (12,652 grams), with a maximum of 38 pounds; the average of four females was 22.5 pounds (10,249 grams), with a maximum of 24.5 pounds. Banko (1960) reported that the minimum weight of eight males at least two years old was 20 pounds, while the minimum weight of 14 females of similar age was 16 pounds. Eight males at least one year old had a minimum weight of 18 pounds, and four females of this age had a minimum weight of 15 pounds. Scott and the Wildfowl Trust (1972) reported the average weight of ten males as 11.9 kilograms, with a range of 9.1–12.5; seven females averaged 9.4 kilograms, with a range of 7.3–10.2. Hansen *et al.* (1971) also presented weight data indicating that ten adult males averaged 11.97 kilograms (range 9.5–13.6), and 11 adult females averaged 9.63 kilograms (range 9.1–10.4).

Identification and field marks. As noted in the whistling swan account, the dorsal surface of the sternum should be examined to be absolutely certain of species identification; the presence of a dorsal protrusion near the sternum's anterior end is the best criterion of a trumpeter swan. Furthermore, if the bird weighs more than 20 pounds (18 if less than two years old), measures at least 50 mm. from the tip of the bill to the anterior end of the nostril, and has entirely black lores or at most a pale yellow or gray mark on the lores, it is most probably a trumpeter swan. Internal examination must be used for determining sex, since there are no known external sexual differences.

In the Field: In the field, the absence of definite yellow coloration on the lores and a voice that is sonorous and hornlike, often sounding like *ko-hoh,* rather than higher pitched and sounding like a barking *wow, wow-wow,* are the most reliable field marks for trumpeter swans (Banko, 1960). Hansen *et al.* (1971) stated that a more nearly straight culmen (the profile of the dorsal edge of the upper mandible) typical of this species, as compared with a more concave culmen in the tundra swan, also provides a useful clue for field identification.

The grayish plumage of the juvenile is held during most of the first year of life, and the lores are likewise feathered for the first few months of life. Second-year birds thus may perhaps be distinguished

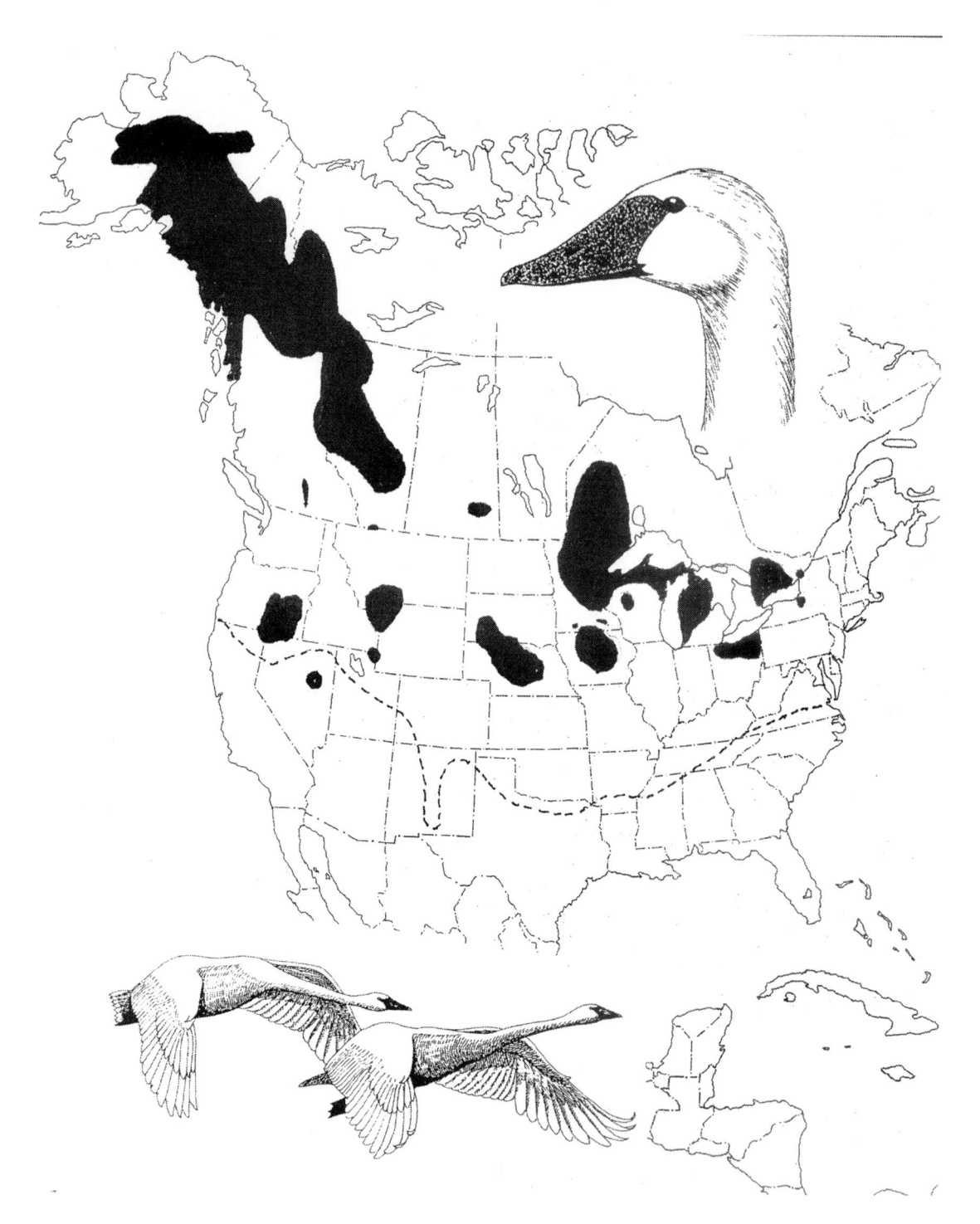

Map 5. Native and reintroduced (as of about 2010) breeding ranges (inked) of the trumpeter swan The approximate southern limits of wintering are indicated by a dashed line. (Mostly after Baldassarre, 2014).

from older ones on the basis of their incompletely developed sexual structures. Young birds have their forehead feathers extending forward to a point on the culmen, while in adults the feathers on the forehead have a more rounded anterior border. Although the birds are usually pure white at the age of 12–13 months, a few dark feathers may persist somewhat longer (Hansen *et al*, 1971).

NATURAL HISTORY

Habitat and foods. Although small cygnets rely on high-protein foods such as aquatic insects and crustaceans, they progressively shift to a vegetable diet as they grow older. Banko (1960) summarized data on trumpeter swan foods and reported use of foliage and tubers of pondweeds *(Potomogeton)*, water milfoil *(Myriophyllum)* leaves and stems, pond lily *(Nuphar)* seeds and leaves, water buttercup *(Ranunculus)* leaves, and a variety of additional herbaceous foods such as *Chara, Anacharis, Lemna, Scirpus, Sparganium, Carex,* and *Sagittaria.* When feeding in shallow waters, trumpeters use their strong legs and large feet to excavate the tubers and rhizomes of various aquatic plants, often forming large holes on the shallow bottoms of the Red Rock Lakes marshes. They also swim with the neck and head under water, pulling rooted materials off the bottom of the ponds. They are also readily able to remove duckweed *(Lemna)* or other small foods from the water surface by straining it through the bill in the manner of dabbling ducks and may feed heavily on duckweed when it is available. Vos (1964) described parental "puddling," a characteristic rapid paddling of feet during swimming, apparently serving to stir food up from the pond bottom. This he observed mostly in an adult female, occasionally in its mate, and several times in a cygnet. Female swans of various species frequently perform this behavior when leading broods, apparently thus improving the foraging efficiency of the short-necked and weak-legged cygnets

Although the trumpeter swan was once strongly migratory, the remaining flocks are now relatively sedentary, with the Canadian or Alaskan population undergoing limited migrations to southeastern Alaska and the western parts of British Columbia (Banko, 1960). Mackay (1957) concluded that swans breeding in the Peace River district of Alberta migrate to the northern United States and mix with swans from the Red Rock Lakes National Wildlife Refuge during winter months, while the breeding areas of those wintering in western British Columbia were then still unknown. Hansen *et al.* (1971) confirmed that these birds represent the Alaskan breeding population.

Banko considered the presence of permanently open water with associated aquatic vegetation, a certain amount of level and open terrain, and a minimum of heavy timber near watercourses as important features of winter habitat. The breeding habitat found in Red Rock Lakes National Wildlife Refuge were characterized by Banko as large shallow marshes or shallow (to four feet deep) lakes, of high fertility, with a profusion of aquatic plants of submerged and emergent growth forms, and with generally non-timbered but well-vegetated shorelines. Within Yellowstone National Park the breeding lakes are generally deeper, more heavily timbered, higher in elevation, and represent more marginal breeding habitat. During the years 1954 to 1957 in Red Rock Lakes National Wildlife Refuge an average of 13 nesting pairs occupied Upper Lake (2,880 acres), 51 occupied River Marsh (8,000 acres), and 15.5 occupied Swan Lake (400 acres), a total average population of about 80 pairs on 11,280 acres, or 4.5 pairs per square mile. Besides the Red Rock Lakes–Yellowstone–Grand Teton population, other major nesting populations occur in Canada and Alaska. Marshall (1968) reported that the nesting population at Grande Prairie, Alberta, numbers about 100 birds, and in Alaska. the birds nest commonly along the southern coast from Yakutat to Cordova and in the Copper River drainage. Additional Alaskan breeding grounds are in the Kantashna, Tanana, Susitna, and Koyukak river valleys, the vast Yukon River delta, the Kenai Peninsula, and the adjacent coast west of the Cook Inlet.

The total 1970 Alaska population was estimated at 2,800 swans, which, added to the Canadian population and an estimated 800 birds in the contiguous United States, may have represented 4,000

Fig. 8. Trumpeter swan, adult in flight

to 5,000 birds (Denson, 1970, Hansen *et al.*, 1971). Although the trumpeter swan is not known to occur in the Aleutian Islands, the whooper swan has been reported there several times (Byrd *et al.*, 1974).

The U.S. Fish and Wildlife Service and Canadian Wildlife Service now recognize three populations of trumpeter swans. The Pacific Coast Population and the Rocky Mountain Population are largely natural. The Interior Population is composed entirely of restored breeding flocks. Transplants from Red Rock Lakes National Wildlife Refuge to other refuges have produced new breeding populations in the coterminous United States. Swans were introduced in Malheur National Wildlife Refuge in Oregon in 1939 and again in 1955, with the first successful breeding in 1958. That same year success occurred in Ruby Lake National Wildlife Refuge, Nevada, after releases in 1949. In 1960 birds were released in Lacreek National Wildlife Refuge in South Dakota, with the first successful nesting in 1963 (Monnie, 1966; Marshall, 1968).

Later introductions were made at the Turnbull National Wildlife Refuge (N.W.R.), Washington, and in the Hennepin County Park District, Minneapolis, Minnesota. After nesting unsuccessfully in 1965 at Turnbull Refuge, later attempts were more successful, and in 1970 a total of eight pairs nested, hatching sixteen young. A few years later the number of swans declined and no nesting occurred for many years. Very recently swans began nesting again at Turnbull. In 2015 there were just four adult swans at Malheur and one cygnet. There were only two adults at Ruby Lake. Restoration efforts in Montana

Trumpeter swan, adult male head profile

sponsored by the Confederated Salish and Kootenai Tribes and the Blackfoot Challenge Partnership are faring much better (Olson, 2016). Besides these refuge nestings, other localized nestings were later reported in Alberta, Saskatchewan, British Columbia, Montana and Idaho. Another Rocky Mountain nesting flock of trumpeter swans consists of the Canadian Flock that breeds in Yukon and Northwest Territories to Alberta and Saskatchewan, the U. S. Breeding Segment (a subunit of the overall Rocky Mountain population) in Idaho, Montana and Wyoming, which had 642 birds in 1954. After a slow decline (there were only two pairs present in Yellowstone National Park in 2015, versus 59 in 1968) these flocks have recovered with the assistance of restoration efforts in Montana, Idaho, and Oregon. A total of 983 swans were found during the fall of 2015 in the U.S. segment of the Rocky Mountain Population, a 61 percent increase over the previous year's survey (Olsen, 2016). (Additions and edits courtesy John Cornely and The Trumpeter Swan Society).

Since 1963 trumpeter swans have bred at Lacreek National Wildlife Refuge, after having been introduced beginning in 1960 from Red Rock Lakes National Wildlife Refuge, Montana. They have since expanded into the Nebraska Sandhills (Ducey, 1999). Nesting has since occurred in many of Nebraska's Sandhills wetlands, especially in Cherry and Grant counties, and was later reported from marshes in Arthur, Brown, Garden, McPherson and Sheridan counties. By 1987 the South Dakota population had increased to at least 268 birds, and by 1995 the Nebraska population had reached about 150 birds. In recent years the swans nesting in South Dakota have declined significantly and most of this flock now nests and winters in Nebraska. In more recent years breeding sites have expanded to include some locations outside the Sandhills region, and Nebraska's Audubon Christmas Count numbers have gradually increased. Wintering is now common on the Snake, North Loup and North Platte rivers, with flocks of 100 or more birds sometimes reported.

Social behavior. Only during the winter season are trumpeter swans appreciably social, and then the limited areas of open water force a degree of sociality upon them. Banko (1960) noted that it is seldom that more than six or eight swans fly together in local flights unless they are simultaneously flushed. He included a photo of 80 birds occupying a small spring in mid-January, but mentioned that as early as February pairs and small flocks begin to spread out over the snowfields that overlie their breeding habitat. As noted earlier, the average refuge density between 1954 and 1957 was 4.5 pairs per square mile (142 acres per pair) in three major nesting habitats, and in the most favorable nesting habitats about 70 acres per nesting pair was recorded during one year. The actual size of the defended area was not determined, but Banko indicated that birds occupying open shoreline usually defended more area than did those nesting on islands, although shoreline nesters sometimes defended only a small bay area around the nesting site. Hansen *et al.* (1971) suggested that spatial isolation, rather than food supply or size of area, was important in determining territorial boundaries.

Trumpeter swans have no significant contact with whistling or mute swans on their breeding or wintering areas, and Banko (1960) reported that they are highly tolerant of other bird and large mammal species. Even among pairs on their breeding territory, the presence of geese, pelicans, cranes, or herons is usually not sufficient to cause aggression, although swans leading young are less tolerant than others. However, one case was found of a nesting swan's killing a muskrat that approached a brood. Vos (1964) also noticed several threats by nesting birds.

Vos (1964) reported on daily activity patterns of three captive swans, which may not be wholly typical of wild birds. He noted that bathing, preening, sleeping, loafing, swimming, and foraging were

Trumpeter swan, pair swimming, male is nearer bird

Trumpeter swan, triumph ceremony

performed several times daily and usually in unison by the pair. Preening bouts typically followed bathing and lasted for varying periods up to 85 minutes. Resting or sleeping followed preening, and favored resting spots were also used for preening and sleeping. Some sleeping periods lasted as long as 85 minutes, and the male usually had longer sleeping bouts than did the female. In total, the adult pair slept about the same amount of time during the egg-laying period, while later in the summer a month-old cygnet slept more than the total of both parents. In general, preening most commonly occurred early in the morning, early in the afternoon, and during the evening. Feeding occurred after the morning and evening preening periods, reaching a maximum in early afternoon, with a secondary evening peak.

There are few good data on daily movements, but Monnie (1966) reported that local movements of up to about a hundred miles were noted at Lacreek National Wildlife Refuge over a prolonged period. Banko (1960) observed that flights during local movements were usually performed at lower altitudes than were longer flights.

Reproductive biology, Monnie (1966) reported that some known-age trumpeter swans (two of nine) initially formed pairs when 20 months old, and initial nesting occurred the following year. Banko (1960) summarized evidence that nesting may begin as early as the fourth year of life or as late as the sixth year, but it would seem probable that these examples are atypical, and that initial nesting in the third year of life would be characteristic. Like mute swans, two-year-old pairs may establish territories, even though actual nesting is not attempted. At least in some cases, the birds may form pairs when 20 months old and begin nesting as early as 33 months after hatching (Monnie, 1966). Some captive swans do not begin nesting until much older, especially if they are reared under wild conditions. A wild-caught pair in the Philadelphia Zoo first nested successfully in 1965, although the female (of unknown age) had been in the zoo since 1959.

Monnie (1966) reported that courtship among 20-month-old swans began in mid-January and continued until mid-March during which time among nine birds two apparent pairs were formed, plus a trio involving two males and a female, while two females remained unpaired. Monnie did not specifically indicate whether this courtship consisted of actual copulatory behavior or of mutual triumph ceremonies. Banko (1960) described the triumph ceremonies of this species, which are typically performed following the expulsion of a territorial intruder. However, he noted that mutual display also regularly occurs in the wintering areas among birds in flocks, although he did not clearly associate this behavior with pair formation. Triumph ceremonies involving more than two birds most probably represent participation by the past season's offspring, if my observations at the Wildfowl Trust are also characteristic of wild birds.

Like other swans, trumpeters are monogamous and have strong pair bonds. Banko (1960) reported a single case of a trio living together, although the sex of the extra bird was not learned. Griswold (1965) also reported a captive trio, in which a male was paired with two females. Banko assumed that a permanent pair bond was typical of this species, although Hansen *et al.* (1971) found one case of a female remating in the year following the loss of her mate.

Vos (1964) observed eleven copulations in captive trumpeter swans, ten of which were seen between April 16 and 26 (the first egg were laid April 21). One copulation was also seen in mid-July, more than a month after hatching had occurred. Typically, as copulation ends both sexes rise together in the water, variably extending their wings (the male usually more fully extending his) while both birds call in unison. Finally, the birds flap their wings once or twice, followed by bathing and preening (Johnsgard, 1965).

Trumpeter swan, triumph ceremony

Trumpeter swan copulation

Banko (1960) reported on 109 nests observed over four seasons in Red Rock Refuge. Over 70 percent of these were located on or very near a previous nest site, with four sites used all four years. Island sites were preferred over shorelines, and fairly straight shorelines tended to be avoided. Highest concentrations occurred where irregular shorelines combined with numerous sedge islands to produce maximum habitat interspersion, producing maximum nest densities of one nest per 70 acres. Hansen *et al.*, (1971) noted that 32 of 35 Alaskan nests were built in water from 12–36 inches deep, and 21 of 40 nests were located in beaver impoundments 6–14 acres in area. Stable water levels and tall, dense emergent plants apparently provide the necessary security, food supply, and nest support needed by these birds.

Most preliminary nest-building is by the female, but the male helps gather nesting material and to a limited extent may assist in nest construction. Females not only spend more time nest-building, but also are more effective in gathering materials (Vos, 1964). Vos did not observe the male actually incubating, but saw it sitting on the nest once during the egg-laying period. However, Griswold (1965) did report an instance of apparent incubation assistance by the male, inasmuch as both birds were once seen on the nest, with four eggs under one and three under the other. This is apparently the only report of possible incubation by the male.

Of 74 completed clutches observed by Banko, the average was 5.1 eggs, with a range of 3–9. Hansen *et al.* (1971) stated that 53 clutches from Alaska's Copper River area averaged 4.9 eggs, while 160 clutches from the Kenai region averaged 5.3 eggs. Yearly differences were noted, with smaller clutches typical of years having late springs, and larger clutches typical of more favorable breeding seasons. The eggs are laid at two-day intervals. Incubation period estimates have ranged from 32–37 days. Hansen *et al.* noted that six nests in the Copper River area had incubation periods of 33–35 days. There is no good evidence that the male assists in incubation under natural conditions.

The cygnets normally hatch at about the same time. However, Griswold (1965) reported a staggered hatching period in one captive pair. He noted that the first two young to hatch were seen entering the water initially when about 48 hours old, while the third left the nest when about 24 hours old. Griswold's observations were complicated by the fact that two females were present, and both may have contributed to the clutch.

Banko (1960) noted that egg-hatching success varied from 51–66 percent during three different years. During six years at Grande Prairie, Alberta, the comparable percentages ranged from 55–92 percent (Mackay, 1964), and three years' data from the Kenai Peninsula, Alaska, indicated an average 82 percent hatching success (Hansen *et al.*, 1971). Infertility and embryonic deaths appear to be the major causes of hatching failure, with egg predation insignificant. A few Alaskan nests have been found destroyed by bears and wolverines *(Gulo luscus)*. Banko (1960) reported one probable instance of renesting following nest destruction.

Vos (1964) noted that for the first few weeks a youngster was closely guarded, with the two parents placing themselves on either side of the cygnet. However, the female was generally more closely associated with it, and usually when swimming the female led the cygnet, with the male following behind. Griswold reported that by the age of about three months a captive female had attained a weight of 14.5 pounds. Four males weighed from 13.5–16 pounds, and collectively averaged about 15 pounds. Banko mentioned a 19-pound cygnet of preflight age, and Hansen *et al.* (1971) stated that such a weight can be attained in as few as 8–10 weeks.

Banko (1960) summarized data indicating that the fledging period is probably normally from 100–120 days, with known minimum and maximums of 91 and 122 days. A very similar range, from

90–105 days, has been reported for Alaskan birds (Hansen *et al.,* 1971). According to Banko, considerable preflight mortality occurs, with possibly 50 percent or more of the young being lost during this period. Most of this mortality occurs early in life, from apparently varied causes.

Monnie (1966) reported cygnet losses by great horned owls (*Bubo virginianus*), and probably also raccoons (*Procyon lotor*), and Banko suspected that minks (*Mustela vison*) or skunks (*Mephitis* spp.) might play a predatory role at the Red Rock Lakes National Wildlife Refuge. Hansen *et al.* (1971) found a rather low (15–20 percent) cygnet mortality rate in Alaska for the first eight weeks, and practically none afterwards.

Wild females have been known to live as long as 18 years, and a male reached 23 years and ten months (Baldassarre, 2014). The apparent maximum longevity record in captivity is 29 years (Scott and the Wildfowl Trust, 1972). Banko (1960) suspected that trumpeter swans are virtually free of most natural enemies once they have fledged and speculated that only coyotes (*Canis latrans*) or golden eagles (*Aquila chrysaetos*) might be of possible significance as predators.

In Alaska, non-breeding birds gather in flocks on large, open lakes and begin their wing molt almost simultaneously, with nearly all of them beginning and terminating their flightless period within ten days of one another. A less regular molting pattern occurs in breeding birds. Males usually begin their wing molt early in the incubation period, or sometimes as late as after hatching. Females begin molting their flight feathers from 7–21 days after the clutch has hatched. Since the flightless period is about 30 days long, both members of a pair are rarely flightless simultaneously, and both sexes regain their flying abilities prior to the fledging of the young.

In Alaska, some young may still be unable to fly at the time of freeze-up, and the birds seem to postpone their fall migration as long as possible, with family groups being the last to leave the breeding

Fig. 9. Trumpeter swan, adult landing

grounds (Hansen *et al.,* 1971). Starvation during severe winters may be a significant mortality factor, at least in Canada. Mackay (1957) mentioned that cygnets of a family evidently remain together for at least the first year after hatching, since three brood-mates that were banded in Alberta in 1955 were all shot in Nebraska the following fall.

Status. Early estimates of this species' population were provided by Hansen *et al.* (1971), who found nearly 3,000 swans in 1968, but probably missed hundreds more during their aerial surveys. There was then thought to be about 150 additional swans that summered in western Canada, and over 300 in the vicinity of Red Rock Lakes National Wildlife Refuge plus Yellowstone and Grand Teton national parks. Of these, 59 birds were concentrated in Yellowstone National Park plus the additional isolated breeding population of about 80 breeding pairs present at Red Rock Lakes National Wildlife Refuge. There were about 600 additional birds in other U.S. refuges and zoos in 1970, suggesting that over 4,000 trumpeter swans then existed in North America.

Trumpeter swans have increased remarkably since the 1970s. As an amazing example of conservation success, the currently expanding Pacific Coast population, which breeds from Alaska to northwestern British Columbia and the Northwest Territories and winters from southern Alaska to Washington and Oregon, numbered 26,790 birds in 2010 (Groves, 2012, Baldassarre, 2014), or seven times the total estimated 1968 North American population! The Interior Population, which once bred across the Great Plains and eastward to the Great Lakes, was totally extirpated during settlement times, but has recently been recreated by transplant programs. These birds now breed from the High Plains to Ontario and New York. By 2010 the High Plains component of this population was centered mostly in Nebraska and South Dakota, and numbered nearly 600 birds.

Other restorations in Minnesota, Ontario, Wisconsin, Michigan, Ohio and Iowa have resulted in nesting flocks. As the Interior Population is expanding, more swans are pioneering south for the winter, and some are congregating in winter flocks in wildlife refuges, sanctuaries, and other areas. Examples include Riverlands Migratory Bird Sanctuary and Squaw Creek National Wildlife Refuge in Missouri, and the Magness Lake area in Arkansas. A few swans are even venturing to their historic wintering areas on the Gulf Coast and Chesapeake Bay. There were nearly 10,000 trumpeter swans in the still rapidly expanding Interior Population by 2010, in widely scattered locations from South Dakota, Nebraska, and Manitoba east to Ontario and New York (additions and edits courtesy John Cornely and The Trumpeter Swan Society). By 2010 the total North American trumpeter swan population was estimated at 46,225 birds, as compared with the 69 that were known to exist in 1932 (Groves, 2012).

Relationships. The trumpeter, whooper, whistling, and Bewick's swans collectively constitute a close-knit evolutionary complex collectively referred to as the northern swans, in which the species limits and internal relationships are still far from clear (Johnsgard, 1974a). Given the conspecificity of the whistling and Bewick's swans, it seems intuitive that the trumpeter swan and whooper swan are each other's nearest relatives (Johnsgard, 1974a), although Livezey (1996) judged the Bewick's swan, rather than the whooper swan, to be the nearest relative of the trumpeter swan.

Suggested readings. Banko, 1960; del Toyo, Elliott & Sargatal, 1992; Mitchell, 1994; Todd, 1996; Kear, 2005; Baldassarre, 2014.

Whooper Swan *Cygnus cygnus* (Linnaeus) 1758

Other vernacular names. None in general English use. Singschwan (German); cygne sauvage (French); cisne gritón (Spanish).

Subspecies and range. No subspecies currently recognized. The breeding range includes Iceland and Eurasia from northern Scandinavia eastward through Finland and northern Russia, and in northern Asia from Kolymsk and Anadyr to Kamchatka and the Commander Islands, and southward through the taiga and scrub forest zone to the Russian Altai, the lower Amur Valley, and Sakhalin. Winters far to the south, from Great Britain and northwestern Europe east to the western border of China. The eastern population also winters in Japan, Korea, as well as along the Pacific coast of China.

Measurements and weights (mainly from Scott and the Wildfowl Trust, 1972). Folded wing: males, 590–640 mm; females, 581–609 mm. Culmen: males, 102–16 mm; females, 97-112 mm. Weights: males, 8.5–12.7 kg (ave. 10.8 kg); females 7.5–8.7 kg (ave. 8.1 kg). Eggs: ave. 113 x 73 mm, white, 330 g.

Identification and field marks. Length 55–65" (140–65 cm). The whooper is the only one of the northern swans with the yellow on its bill reaching forward beyond the nostrils, and the only swan with a partially yellow bill that has a folded wing length in excess of 580 millimeters. In the adult plumage both sexes are entirely white, with black legs and feet and a black bill except for the large yellow bill marking. *Females* cannot be externally distinguished from males. *Juveniles* have a variable number of grayish feathers mixed with the white feathers of the immature plumage, and their bills are pinkish at the base rather than yellow.

In the field, the extensively yellow bill is the best field mark, and the more trumpet-like and less musical call of this species helps to distinguish it from the noticeably smaller Bewick's race of the tundra swan.

NATURAL HISTORY

Habitat and foods. The preferred breeding habitat is shallow fresh-water pools and lakes, and along slowly flowing rivers, primarily in the coniferous forest (taiga) zone, but also in the birch forest zone and treeless plateaus, but rarely in tundra (Voous, 1960). On the breeding grounds the foods of adults probably consist mainly of the leaves, stems, and roots of aquatic plants, including algae. A considerable amount of grazing on shoreline and terrestrial vegetation is performed; the whooper swan is thought to consume higher proportions of terrestrial plants and animal materials than is true of the Bewick's tundra swan. In the wintering areas of Europe the birds have been found to consume such aquatic plants as pondweeds and similar leafy foods *(Zostera, Ruppia, Elodea)* that can readily be reached from the surface. In some areas they also graze on winter wheat, waste grain, turnips, and potatoes (Owen & Kear, in Scott and the Wildfowl Trust, 1972). In some areas animal materials are consumed, such as midges and freshwater mussels (Kear, 2005).

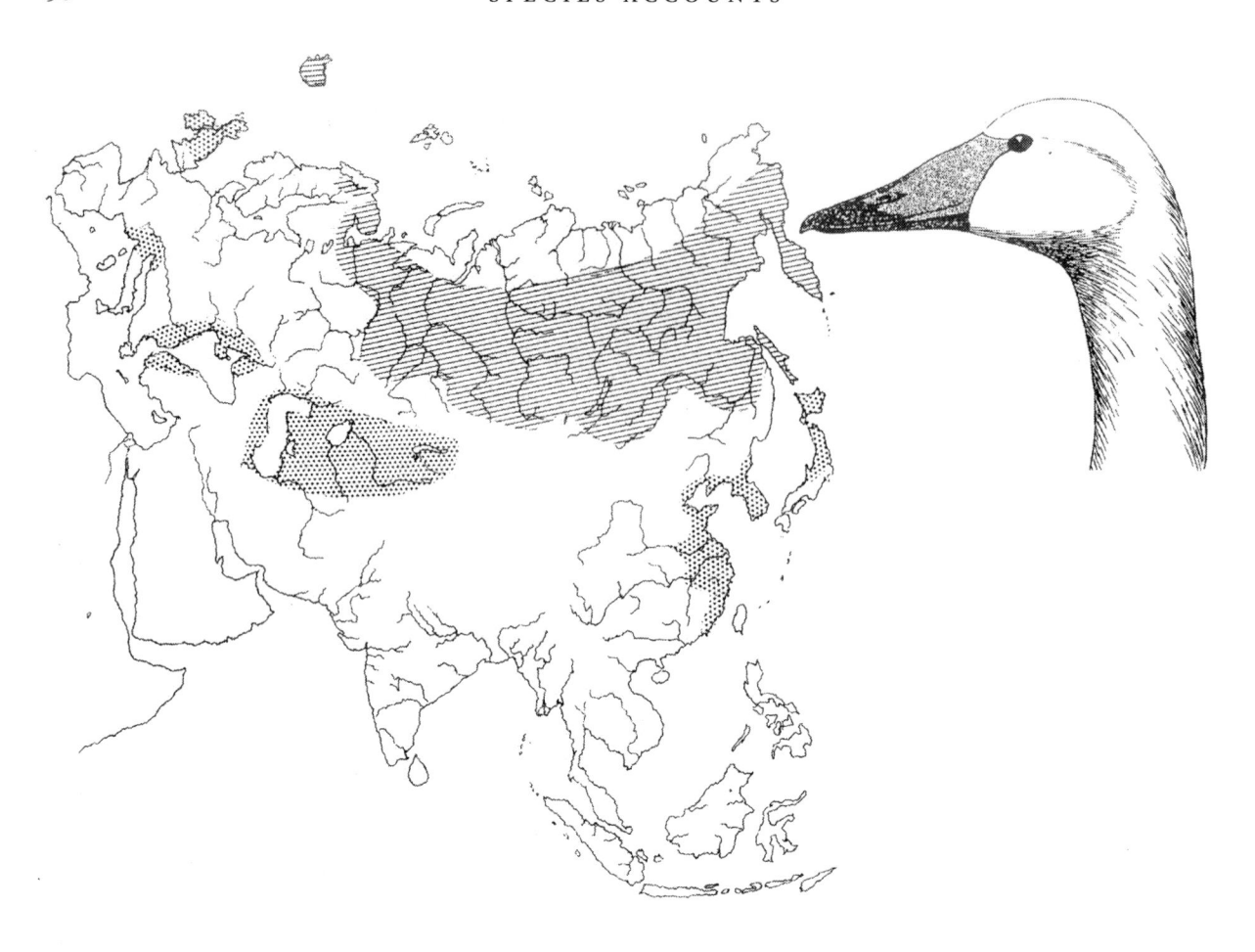

Map 6. Breeding (hatched) and wintering (stippling) distributions of the whooper swan (from Johnsgard, 1978).

Social behavior. Whooper swans are gregarious and form large flocks in the nonbreeding season; in Aberdeenshire, Scotland, flocks of 300–400 often may be seen foraging in agricultural fields. Fairly large winter flocks also occur locally in Japan in protected sites such as Hyoko reservoir (Lake Hyo), on the island of Honshu. These wintering flocks consist of firmly paired, courting, and sexually immature birds, including first-year birds still consorting with their parents.

So far as is known, the process of pair formation and pair bonding in whooper swans is exactly as in the trumpeter swan, as is the behavior associated with copulation. First-year birds remain with their parents through the winter period and start back toward the breeding grounds with them. Since the birds arrive on the breeding grounds in pairs, the family bonds must be broken during the spring migration period. Whooper swans leave their European and Asian wintering grounds in late February or March, and may not arrive at their northernmost breeding areas until May.

Although essentially monogamous, divorce rates are much higher in whooper swans than in Bewick's swans (Rees *et al.* 1996). In one study 5.8 percent of the whooper pairs were found to have a new mate although their earlier mate was still alive. Divorce is not obviously associated with poor breeding success, since 25 percent of divorced pairs had bred the previous year. However, breeding success improved among pairs remaining together for several years.

Fig. 10. Whooper swan, adults landing

Fig. 11. Whooper swan, pair performing triumph ceremony

Reproductive biology. Immediately upon the birds' arrival on their breeding grounds, or at most within two weeks of arrival, nest building begins. Nests are built either on dry ground or in reed beds, often so large and with so deep a cup that the top of the sitting female may be flush with the rim of the nest. The nest cup is extensively lined with down, and a clutch of from 4–7 eggs, most commonly five, is laid. Average clutch sizes in Iceland and Finland are 4.5 and 4.4 respectively, with clutches in southern Finland slightly larger (and cygnet morality lower) than in northern Finland (Einarsson, 1996).

Eggs are laid at approximate 48-hour intervals. In Russia, egg-laying occurs during May and June, and in Iceland it also normally occurs at that time. Undoubtedly the time at which the nesting sites become snow-free dictates the year-to-year onset of nesting in this species. The probable usual incubation period is 35 days, although various estimates have ranged widely, from 31–42 days.

Whooper swan post-copulatory display, male with wings lifted

Incubation is performed entirely by the female, but the male remains in very close attendance. Typically she leaves for a short time during the warmest part of each day to forage. Some cases of second clutches following clutch loss have been reported. Hatching of the clutch is synchronous, over a 36–48-hour period. The male also closely guards the cygnets, which in Iceland have been reported to fledge in as short a period as about two months, but is typically 87 days (Haapenen *et al.*, 1973b). Captive-raised young may fledge in as little as 80 days (Kear, 2005). The young birds eat insect larvae, adult insects, and vegetation growing on the water surface or just below it. The juveniles remain with their parents through their first winter. Maximum longevity record in captivity is 25 years (Scott & the Wildfowl Trust, 1972).

Status. The total world population of the whooper swan was less than 100,000 birds in about 1970, but nearly universal protection has been given the species. The Icelandic breeding component was then 5,000–6,000 birds, which wintered in Iceland and Great Britain. Those breeding in Scandinavia and western Russia and wintering in northwestern Europe probably numbered about 14,000. The birds that breed farther east in Russia wintered mainly on the Black and Caspian seas, and probably then totaled at least 25,000. The Far Eastern breeding component that winters along the western Pacific coast is least well documented, but close to 11,000 were counted during a wintering census in Japan in that same time period (Scott and the Wildfowl Trust, 1972).

Whooper swan, female with downy cygnets

In more recent surveys, the Icelandic population was judged to be nearly 21,000 in a 2000 survey, with 30–33 percent wintering in Great Britain, 61–66 percent in Ireland, and the rest wintering in Iceland (Kear, 2005. The Northwestern European population was surveyed in January, 1995, with over 20,000 birds judged to present in Denmark, 14,000 in Germany, 7,500 in Sweden, over 5,000 in Norway, and 3,000 in Poland. The total Northwestern European population was estimated at 59,000 (Laubek *et al.,* 1999). The species' eastern population is dispersed across Russia and northern China, with those wintering in western Asia around the Black and Caspian seas of unknown population size, but possibly numbering about 20,000. Those breeding from eastern Russia through China, and wintering in China, Japan and Korea are likewise of uncertain population size, but during the 1990s they were estimated at about 60,000 in China and 31,000 in Japan. Perhaps another 4,000 birds winter on the Korean Peninsula. Small numbers also winter on the Aleutian and Pribilof islands (Kear, 2005).

Relationships. Although the whooper swan and trumpeter swan have at times been regarded as conspecific (e.g., Delacour, 1954-65), this is not the only possible interpretation of relationships among the northern swans (Johnsgard, 1974), and the usual present position is to consider the two as separate species. For example, Livezey (1996) judged the Bewick's swan, rather than the trumpeter swan, to be the whooper swan's nearest relative.

Suggested readings. Scott and the Wildfowl Trust, 1972; del Toyo, Elliott & Sargatal, 1992; Todd, 1996; Kear, 2005.

Tundra Swan (Whistling Swan) *Cygnus c. columbianus* (Ord) 1815

Other vernacular names. wild swan, whistler, Pfeifschwan (German); cygne siffleur (French), cisne silbador (Spanish).

Range: Breeds in arctic North America from western Alaska across the northern parts of the Northwest Territories to Southampton Island, Nottingham Island, and the Belcher Islands. The North American population winters mostly along the Atlantic and Pacific coasts, but passes through the interior during migrations, and varying numbers overwinter in northern Utah.

Subspecies. The whistling swan is currently classified by the American Ornithologists' Union (1998), Kear (2005) and Baldassarre (2014) as a North American subspecies of *C. columbianus*, the tundra swan. This species, which includes the Eurasian Bewick's swan (*C. columbianus bewickii*), has a high-latitude Holarctic distribution, the North American component breeding in arctic tundra habitats from western Alaska to Hudson Bay, and on Southampton, Banks, Victoria, and St. Lawrence Island. Winters along the Pacific coast from southern Alaska to California, with some overwintering inland to Utah; also winters on the Atlantic coast, mainly from Chesapeake Bay to Currituck Sound, but increasingly north to northern New England.

Measurements and weights (after Banko, 1960). Folded wing: Adult male 501–569 mm. (average 538), adult female 505–561 mm. (average 531.6). Culmen: Adult male 97–107 mm. (average 102.6), adult female 92.5–106 mm. (average 99.9). Weights: males, 4.7–9.6 kg (ave. 7.1 kg); females, 4.3–8.2 kg (ave. 6.2 kg). Nelson and Martin (1953) indicated an average weight of 35 males as 15.8 pounds (7,165 grams), with a maximum of 18.6 pounds; 42 females averaged 13.6 pounds (6,167 grams), with a maximum of 18.3 pounds. Banko (1960) reported that seven males at least two years of age had a maximum weight of 19.5 pounds, and 21 females of the same age class had a maximum weight of 19 pounds. Sherwood (1960) mentioned a male that weighed 19 5/8th pounds. Scott *et al.* (1972) reported the average weight of 29 males as 7.5 kilograms (range 7.4–8.8) and 39 females averaged 6.6 kilograms (range 5.6–8.6). Eggs: ave. 110 x 73 mm, white, 280 g.

Identification and field marks. Length 48–58"(120–50 cm). Whistling swans can be confused with trumpeter swans even when being handled; the absence of a fleshy knob at the base of the bill readily separates them from mute swans. To be certain of identification, the upper surface of the sternum must be examined to see if a protrusion near its anterior end is present, which would indicate a trumpeter swan. Alternatively, the bird is probably a whistling swan if it weighs less than 20 pounds, measures less than 50 mm. from the tip of the bill to the anterior end of the nostril, and has bright yellow or orange yellow spots on the lores.

The whistling swan is completely white in *adult* plumage, with black legs and feet and a bill that is typically entirely black except for a small yellow area in front of the eye. However, whistling swans

Map 7. Breeding (hatched) and wintering (stippling) distributions of the whistling race of the tundra swan (from Johnsgard, 1978).

sometimes lack this yellow mark, and thus a bill length that is less than 50 millimeters from the front of the nostrils to the tip of the bill is a better criterion for birds in the hand. *Females* are identical to males, and average slightly smaller in size. *Juveniles* possess some gray feathers for most of their first fall and winter of life, and their bills are mostly pinkish.

In the field: The neck of the whistling swan appears to be shorter and the bill profile somewhat more concave than those of the trumpeter swan. Unless both trumpeter and whistling swans are seen together, a size criterion is of little value in the field. Rather, the differences in their voices are perhaps the best field mark, in association with the presence or absence of yellow coloration on the lores. If the lores are completely black, the bird may be of either species, but if a prominent yellow to orange yellow mark is present, the bird is a whistling swan. Further, if the voice is sonorous and hornlike, often sounding like *ko-hoh*, it is a trumpeter, whereas the voice of the whistling swan is more like a high-pitched barking sound, *wow, wow-wow* (Banko, 1960).

No external differences in the sexes exist that would allow for sex determination without internal examination. Birds possessing feathered lores and/or some grayish feathers persisting from the juvenile plumage are in their first year of life. Apparently the rate of sternal penetration of the trachea is fairly constant for the first three years, and by the second winter the tracheal loop starts to rotate and begin its expansion into the carina of the sternum (Tate, 1966). Together with the length of the tracheal perimeter within the sternum, the changes in the shape of the nasal bones are good indicators of age, according to Tate. First-year birds have a well-defined "V" groove formed by the nasal and lachrymal bones, which gradually alters by medial fusion with age, so that the V is nearly obliterated in old birds. In young birds the feathers of the forehead extend forward to a point in the midline, while in older birds this point gradually recedes until a smooth and rounded brow is formed.

NATURAL HISTORY

Habitat and foods. The whistling swan is associated with arctic tundra throughout its breeding range in North America, and thus is an ecological counterpart of the Old World subspecies, the Bewick's swan. The whistling swan has a breeding range well to the north of the trumpeter swan's, in arctic tundra. Heaviest nesting concentrations in Canada are in the coastal strip from the west side of the Mackenzie Delta to the east side of the Anderson Delta, with sparser populations inland, especially south of the tree line (Banko and Mackay, 1964). This Northwest Territories population evidently winters on the Atlantic coast (Sladen and Cochran, 1969). In central and eastern Canada swans are usually absent from the rocky Precambrian shield, but occur wherever typical tundra occurs, north to Banks Island and south to about the Thelon River. In Alaska, major breeding areas include the north side of the Alaska Peninsula and adjoining Bristol Bay, the Yukon–Kuskokwim Delta, and, to a much lesser extent, the Kotzebue Sound area

Whistling swans winter in two widely separated areas. Approximately half the continental population winters in the Atlantic Flyway, primarily on Chesapeake Bay and Currituck Sound. The rest of the population winters in the Pacific Flyway, chiefly in the Central Valley of California. Some usually also overwinter in the Great Salt Lake valley of Utah, where the yearly numbers there are influenced by the severity of the winters (Sherwood, 1960). Normally their winter habitat includes sufficient aquatic

Fig. 12. Whistling (tundra) swan, pair in flight

plant life to provide adequate food, but during unusually severe winter conditions feeding in cornfields has been observed (Nagel, 1965).

Preferred wintering habitat in the Chesapeake Bay area consists of open and extensive areas of brackish water no more than five feet deep (Stewart, 1962). January counts in that region indicated the following percentage usage of available habitats: brackish estuarine bays, 76 percent; salt estuarine bays, 9 percent; fresh estuarine bays, 8 percent; slightly brackish estuarine bays, 6 percent; and other habitats, 1 percent. Freshwater areas are used primarily by early fall arrivals.

Foods taken on the breeding grounds are not yet well studied, but in migration and wintering areas the birds usually feed extensively on such aquatic plants as wild celery *(Vallisneria),* wigeon grass, bulrushes, and pondweeds. The tubers of arrowhead *(Sagittaria)* are favored foods, and in brackish waters the birds may feed to some extent on mollusks, especially clams. Some grazing on agricultural lands is performed by wintering birds, which may consume grain and waste potatoes (Owen & Kear, in Scott and the Wildfowl Trust, 1972).

Like other swans, the whistling swan feeds predominantly on vegetable materials from aquatic plants. Martin *et al.* (2011) listed grasses and sago pondweed (*Potamogeton pectinatus*) as major food for both the eastern and western populations, and additionally list wild celery (*Vallisneria*), lady's thumb (*Polygonum persicaria*), horsetail (*Equisetum*), and bur reed (*Sparganium*) as important foods in one region or the other. Sherwood (1960) reported that tubers and seeds of sago pondweed were the exclusive food of twelve specimens obtained in the Great Salt Lake valley, although other aquatic foods were available.

Since these swans typically feed on or closely adjacent to their nesting areas, they normally are not forced to move about extensively in search of food. Thompson and Lyons (1964) noted that pronounced diurnal foraging flights were not characteristic of the spring flock of whistling swans they studied and noted that average midday counts were only about 200 birds fewer than average morning or evening counts (749 and 771, respectively). Sladen and Cochran (1969) observed that swans rarely reached an altitude of 1,000 feet during local movements.

Edwards (1966) noted the presence of wintering whistling swans in the flock of resident mute swans at Grand Traverse Bay, Michigan. Martin *et al.* (1951) and others have suggested that whistling swans may despoil the supply of duck foods in some areas, and certainly the preferred foods such as sago pondweed and wigeon grass are also used by many ducks. Wigeons and canvasbacks are species with habitat preferences and foods similar to those of whistling swans in the Chesapeake Bay region (Stewart, 1962). Sherwood (1960) mentions observing a considerable number of species of geese and swans feeding among swans without any visible intolerance on the swans' part. He suggested that the swans may actually increase the forage for the ducks, both by pulling up more food than they actually consume and by possibly creating new sago beds by dissemination of seeds and tubers as well as by "cultivation" of the marsh bottom.

Stewart and Manning (1958) and Stewart (1962) reported on the winter foods of whistling swans in Chesapeake Bay and found that birds foraging in the preferred brackish estuarine bay habitat relied largely on wigeon grass (*Ruppia*) and to a lesser extent on sago pondweeds, with bivalve mollusks (*Mya* and *Macoma*) also being taken in considerable amounts. Four birds collected in fresh water estuaries had been feeding almost exclusively on wild celery, and four from estuarine marsh ponds had been eating wigeon grass, three-square (*Scirpus*), and grasses.

Whistling (tundra) swan, adult swimming

Whistling (tundra) swan, adult swimming, front view

Social behavior. Flock sizes in wintering areas and during migration are often large, and may number in the hundreds or even in the thousands, although the birds are strongly territorial and well scattered during the breeding season. The birds have a relatively long migratory route, often of more than 2,000 miles between wintering and breeding grounds. Counts made during spring in Wisconsin indicate that most flocks consist of units of up to as many as about 13 birds that remain together on local foraging flights. Thus, families and pairs are the obvious unit of substructure in whistling swan flocks. By the time they reach their breeding grounds these flocks have broken up and the birds spread out widely over the tundra, often in densities of only about one or two pairs per square mile. Pair formation probably occurs in wintering flocks, presumably when the birds are in their second or third winter.

Very little reliable information is available on the age of sexual maturity in whistling swans. They have been bred only rarely in captivity; Delacour (1954) reported breeding by a five-year-old female with an older male, and Robert Elgas (pers. comm.) successfully bred a pair of hand-reared whistling swans when they were six years old. Two pairs of swans hatched from wild-taken eggs nested initially when they were four years old (William Carrick, pers. comm.). Scott (1972) believed that the conspecific Bewick's swan may normally breed initially at four years. There seems to be no record of captive whistling swans breeding before their fourth year, but it is uncertain that this age is typical of wild birds. The triumph ceremonies and behavior associated with copulation are nearly the same in this species as in the trumpeter swan, although the speed of movements and associated vocalizations differ appreciably (Johnsgard, 1965).

During the nonbreeding season whistling swans are highly social, with flock sizes often numbering in the hundreds. Thompson and Lyons (1962) made observations on a flock of 1,022 swans during spring migration in Wisconsin and counted the birds in groups making local movements to and from foraging areas, mostly on fallow fields nearby. Nearly 35 percent of the flock counts were of pairs, with units of 3–5 birds also fairly common. This would suggest that yearling birds often remain with their parents during spring migration, although no attempt was made to distinguish young birds from adults. Apart from a small percentage of single birds, the remaining flock sizes gradually diminished in frequency up to a unit size of 13 birds. Among the Bewick's swans wintering at the Wildfowl Trust, up to three seasons' of young have been observed associating with their parents, producing flock units of 13–15 birds. Thus, it is apparent that even large flocks of swans have a well-developed substructure that is related to family bonding.

Whistling swans are to be found in flocks consisting of aggregated pairs and family groups at all times except during the nesting season. Such groups often merge in "staging areas" at various points along their migration routes; these areas provide a combination of abundant food and relative safety from large predators. Fall flocks of 10,000–25,000 swans have been reported in Alberta and Utah (Banko and Mackay, 1964). Staging areas often consist of temporarily flooded fields or permanent water areas no more than about five feet deep.

The low densities of swans on their breeding grounds is probably a reflection of territorial tendencies. Lensick (1968) reported nesting densities of from 130–320 hectares per pair (0.8–2.0 pairs per square mile) in western Alaska. Swan densities based on aerial surveys in the Northwest Territories were estimated from 1948 through 1953. In the wooded delta of the Mackenzie River densities averaged 1.5 swans per square mile. In the area between the Mackenzie and Anderson rivers, the comparable averages were: coastal tundra, 1.7; upland tundra, 1.3; and transition zone to coniferous forest, 0.3 swans

Whistling (tundra) swan, adult head profile.

per square mile (U.S. Fish and Wildlife Service, *Special Scientific Report: Wildlife* –No. 25). In 1950 the area from the Annak River to Kent Peninsula was also surveyed and found to have a swan density of 0.16 per square mile, while southwestern and southeastern Victoria Island had a much lower density of only 0.007. It would seem that a maximum density of about one or two pairs per square mile might be expected in very favorable lowland tundra habitats.

Reproductive biology. Little is known of the pair-forming behavior of whistling swans, but it is probably comparable to that of the better-studied Bewick's subspecies. Peter Scott (1966) noted that two-year-old birds spent quite a lot of time in courtship display during the winter months. However, Dafila Scott (1967) mentioned that many of the pair bonds formed during second-winter birds were only temporary and usually were broken by the following winter. As with the other swans, pair formation is a gradual and inconspicuous process, with a major aspect being the tendency of males to defend mates or potential mates and, after expelling intruders, to return to the female, where they join in a mutual triumph ceremony (Johnsgard, 1965). Differences in the head shape and bill patterning are apparently important bases for individual recognition among the arctic-breeding swans, and it is highly probable that individual differences in vocalizations may also play a role in mate recognition.

Like the trumpeter swan, copulation in whistling and Bewick's swans is preceded by mutual head-dipping movements that closely resemble those of bathing birds. Unlike the mute swan, preening movements do not playa role in precopulatory behavior. As treading is terminated, the male releases his grip on the female's nape as both birds extend their necks strongly upward and utter loud notes, usually simultaneously extending and waving their wings (Johnsgard, 1965).

Arrival on the nesting grounds of western Alaska occurs in late May, and nesting is usually underway by the first of June. There is a high degree of synchrony of nest initiation in individual areas, so that most hatching occurs over a period of only three or four days. Most pairs choose nest sites on the shore of a lake or a pond within 20 yards of water; somewhat fewer nests are built on islands or points of lakes, and even fewer on tundra or in other locations. Elevated sites, such as hummocks, are favorite nest sites and are also among the first areas to be snow-free in spring. There are marked year-to-year differences in average clutch sizes, but they usually range from 3–5 eggs, with smaller average clutch sizes associated with late, cold springs. Incubation is normally performed by the female, although a few observers have seen males sitting on and apparently incubating the eggs.

The nests of whistling swans are usually mounds of moss, grasses, or sedges and are from one to two feet high (Banko and Mackay, 1964). Nests of whistling swans are typically well scattered over the tundra. Banko and Mackay (1964) reported that nest sites vary in location from the edge of water to the top of low hills a half-mile from water, with small islands in tundra ponds being preferred locations.

The female usually assumes all the incubation duties, as with other white swans, but the male remains close by and actively guards the nest. Egg-laying begins shortly after arrival at the tundra breeding grounds in late May or early June, and hatching occurs in late June or early July (Banko and Mackay, 1964). In southeastern Victoria Island, at the northern edge of this subspecies' species' range, the nests are constructed in as little as five days or less, and in one case a nest was built and three eggs were deposited in no more than eight days (Parmelee *et al.*, 1967). According to Banko and Mackay (1964), four eggs constitute the normal clutch, with as many as seven at times. Lensick (1968, and in Scott *et al.*, 1971) reported that five eggs were the normal clutch size in good springs, but only three or four eggs usually present in cold, wet springs. The average clutch size of 297 clutches was 4.3, with a mode of five and a range of 1–7.

An incubation period of 30–32 days is typical; Banko and Mackay (1964) estimated the whistling swan's average incubation to last about 32 days. Robert Elgas (pers. comm.) noted a 30-day incubation period for Alaskan whistling swan eggs incubated under geese. Hatching there begins in early July, and young are probably still about into September. No doubt a critical relationship exists between the time of fledging and the first freezing weather, which may greatly influence breeding success during some years.

Most broods remain within 100–400 yards of their nest for some time, and the young grow at a very rapid rate, reaching weights of 11–12 pounds in 70 days, and fledging at about the same time, at 60–75 days of age. Banko and Mackay reported that hatching occurs in late June or early July, while fledging occurs about the middle of September, suggesting an approximate 75–80-day fledging period. In Alaska the adults undergo their flightless period in July and early August, and non-breeders begin to regain their power of flight in late August. About 85 days after the peak of hatching, families with fledged young begin to join the non-breeders for the fall flight southward (Bellrose, 1976).

Little quantitative information is available on hatching success, but Banko and Mackay (1964) estimated that an average of only two or three cygnets per hatched clutch survived until fledging in

Fig. 13. Whistling (tundra) swan, adult landing

autumn. By counting the percentage of the distinctively plumaged juveniles during fall and winter, estimates of productivity and mortality can be attained. Chamberlain (1967) noted that the percentage of young birds in the 1964-1965 winter flocks on Chesapeake Bay ranged from 9.46–13.9 percent, while in 1965-1966 they ranged from 8.22–12.1 percent. The percentage of young was highest during January counts because of the relatively later arrival of family groups than of non-breeders.

For comparison, average brood sizes typically range from 2.55–2.63 young per pair in Alaska and the Northwest Territories. Winter brood counts have ranged from 2.15–2.63 cygnets/pair, suggesting a cygnet mortality of about 18–25 percent. During the eight-year period between 1964 and 1971, the percentage of juveniles in the Atlantic coast wintering population ranged from 4.8–14.6 percent (ave. 11.1) and the average number of cygnets per family varied from 1.54–2.24 (ave. 1.93) birds (J. J. Lynch, in unpublished "Progress Reports of Productivity and Mortality among Geese, Swans, and Brant").

Information on adult mortality rates in whistling swans is sparse, since few are banded and they have been legal game in only a few states. Hunter kills in the eastern population averaged 3,382 birds from 2000–2008, and 1,057 birds in the western population from 1994–2010 (Baldassarre, 2014). Based on sightings of nearly 6,000 neck-banded birds, survival in the eastern population was judged to be 92 percent for adult males and females, 81 percent for juvenile males and 52 percent for juvenile females (Nichols *et al.*, 1992). Wild females have been known to live as long as 23 years (Baldassarre, 2014).

The postnuptial molt of the adults occurs while the young are still flightless, the female becoming flightless about two weeks after the young hatch, while the male molts about the time the female regains her flight (Banko and Mackay, 1964). Assuming each may be flightless for about a month, the adults should both have regained their powers of flight by the time the young are about 80 days old, or nearly fledged themselves.

At that time, or mid-September, a fairly leisurely fall migration southward begins through the interior of northwestern Canada along the Mackenzie River valley. By early October large concentrations of birds occur on Lake Clair and Richardson Lake in northeastern Alberta. When the flocks reach eastern Montana and western North Dakota the combined eastern and western population splits into two groups, depending on whether they will fly southwest across Montana and winter in some of the western states, especially California, or turn southeast, crossing North Dakota and southwestern Minnesota, toward wintering grounds along the mid-Atlantic coastline.

Status. The North American population of whistling swans probably consisted of somewhat less than 100,000 birds in the 1970s, judging from federal surveys. Slightly over half of the total North American population nests in Alaska. There has been an approximate doubling in estimated whistling swan numbers over the past half-century; the western population had increased to more than 105,000 by 2009, and the eastern population to nearly 97,000 by 2010 (Baldassarre, 2014). There have been regular albeit limited and localized hunting seasons for whistling swans since 1962, but the present hunting mortality levels of a few thousand birds have seemingly not had any measurable effects on the population.

Relationships. The likely evolutionary relationships of the northern swans have been discussed earlier (Johnsgard, 1974a). Although the current consensus is that the Bewick's and whistling swans are conspecific races of the tundra swan, there are some conflicting points that do not entirely support this merger, whereas in contrast to some earlier opinions the trumpeter and whooper swans are now considered two distinct species.

Suggested readings. Scott & the Wildfowl Trust, 1972; Bellrose, 1976; del Toyo, Elliott & Sargatal, 1992; Limpert and Earnst, 1994; Todd, 1996; Kear, 2005; Baldassarre, 2014.

Bewick's (front) and whistling (rear) tundra swan adults.

Tundra Swan (Bewick's Swan) *Cygnus columbianus bewickii* Yarrell 1830

Other vernacular names. None in general English use. Zwergschwan (German); cygne de Bewick (French); cisne de Bewick (Spanish). The collective term "tundra swan" is generally used to designate the entire Bewick's swan–whistling swan complex, based on the current evidence that they consist of a single Holarctic species.

Subspecies and range. The Old World breeding range extends from the Pechenga River, near the Fenno-Russian border, eastward along the north Siberian coast to about 160 degrees East Longitude, as well as on Kolguev Island and southern Novaya Zemlya. Besides the North American race, an eastern Siberian race, *jankowskii,* has sometimes been recognized for birds breeding east of the Lena River delta. Winters in two widely separated areas: Europe (especially the Netherlands) and eastern Asia (China, Japan, and Korea).

Measurements and weights (mostly from Scott & the Wildfowl Trust, 1972). Folded wing: males, 485-573 mm; females, 478–543 mm. Culmen: males, 82–108 mm; females, 75–100 mm. Weights: males, 4.9–7.8 kg (ave. 6.4 kg); females, 3.4–6.4 kg (ave. 5.0 kg). Eggs: ave. 103 x 67 mm, white, 260 g.

Identification and field marks. Length 45–55" (115–40 cm). This is the smallest of the northern swans; it is the only one with a partially yellow bill that has a wing measurement of less than 575 mm in adults, and the yellow markings of the bill do not extend below and beyond the nostrils. *Females* are identical to males, and *juveniles* have mottled gray and white plumage for most of their first fall and winter of life. With increasing age, yearling and older birds gradually become indistinguishable from adults.

In the field, the relatively small size and high-pitched musical call of this species provide useful field marks. The extent of yellow on the bill may be useful for close-range identification. It is also substantially smaller than the whooper swan.

NATURAL HISTORY

Habitat and foods. The breeding habitat of this species consists of shallow tundra pools with abundant submerged vegetation and a luxuriant growth of shoreline vegetation (Voous, 1960). Virtually no specific information on breeding-grounds foods is available, but in Russia the birds are reported to consume a variety of aquatic plants and grassy territorial plants. As with the other northern swans, the leaves, roots, and stems of pondweeds are favored foods on migration and in wintering areas, and the roots and rhizomes of eelgrass *(Zostera)* are also an important food. Various pasture grasses *(Glyceria, Agrostris,* etc.) are grazed extensively when they are wet or flooded in England, and

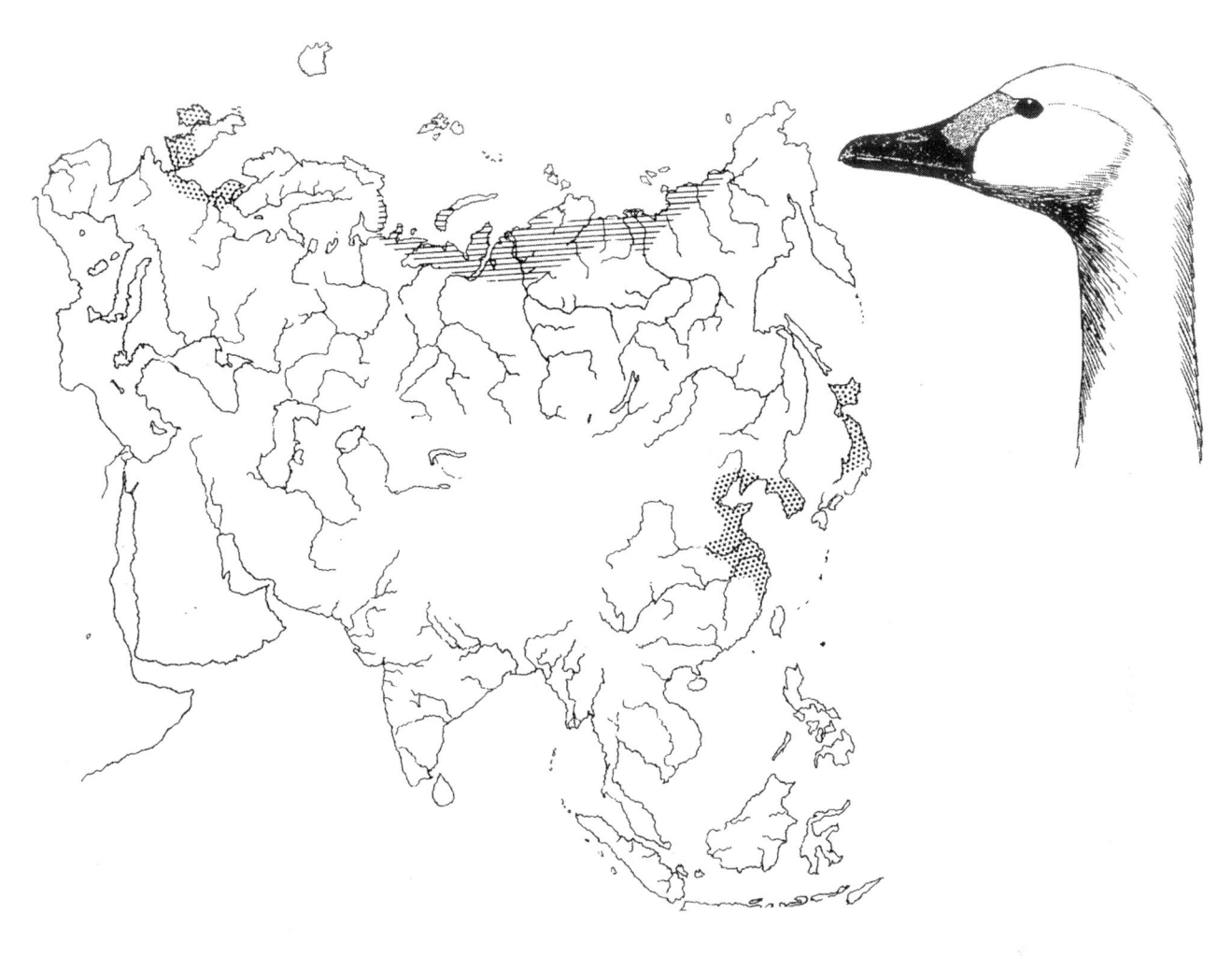

Map 8. Breeding and wintering (stippling) distributions of the Bewick's race of the tundra swan (from Johnsgard, 1978).

sometimes grazing on grasses of drier pastures is also done (Owen & Kear, in Scott and the Wildfowl Trust, 1972). During the early 1960s at the Wildfowl Trust, Bewick's swans typically roosted on the mud flats of the nearby Severn River and few in twice daily to the Trust grounds to eat the grain put out for them. At times they would stay at the Trust all day, returning to the river only after the late afternoon feeding period.

Social behavior. Probably more is known of the social behavior of this swan than any other of the northern swans, as a result of extensive observation of individually identified birds at the Wildfowl Trust (currently known as the Wildfowl and Wetlands Trust) in England. Like other swans, the pair bonds of this species appear to be strong and potentially permanent. Peter Scott (1972) reported that there had been no observed cases of "divorce" among hundreds of individually recognizable Bewick's swans over seven years of observation, and up to three years had been required for bereaved swans to take a new mate. Dafila Scott (1967) reported that some swans have left in the spring with one mate and returned the next fall with a different one, suggesting that mate replacement sometimes occurs during a single breeding season. Some tentative pairing may occur during the second winter, but in six of seven cases she observed, these pairings had broken up by the following winter. Peter Scott (1972) noted, however, that some swans may remain with their parents for their second or even third winter of life.

Through this work it is known that pair bonds are strong and permanent, without a single known case of "divorce" among hundreds of pairs studied. Birds that lose a mate may take up to three seasons to find a new one, but some establish new bonds much sooner. Further, family bonds persist in immature birds even up to the third winter of their lives, resulting in the association of up to four generations of related birds. Some tentative pairing may occur in birds as early as their second winter of life, but nearly all these bonds are broken by the following winter, and probably initial breeding does not occur in this species until the birds are nearly four years old (Scott and the Wildfowl Trust, 1972). As in the other swans, pairs are formed and maintained by mutual displays such as the triumph ceremony; copulation probably plays little if any role in such pair-bond development as it rarely if ever occurs among wintering flocks. Bewick's swans winter at great distances, usually well over a thousand miles, from their breeding grounds, and their departure from England for the breeding grounds occurs during January and February. Arrival on the breeding grounds of Siberia may not occur until May, and is associated with the onset of thaws and the appearance of flowing water in rivers (Dementiev & Gladkov, 1967).

Reproductive biology. These swans spread out greatly on reaching their breeding grounds and select nest sites immediately. These sites are usually small hummocks on the tundra providing good visibility. The nests are constructed of tundra vegetation, especially sedges and mosses, with a lining of down. It is typical that the pair uses an old nest site after some refurbishing, with the female lining the nest with down or other feathers (Dementiev and Gladkov, 1967). The first eggs are laid on Novaya Zemlya at the end of May or early June, and clutch sizes are reportedly only 2–3 eggs there, although elsewhere 3–4 eggs are probably a more typical normal range for the clutch. Other observations have indicated a mean clutch size of 3.62 eggs, with a range of 1–6, with year-to-year variations depending on climate and food conditions (Kear, 2005).

The female alone incubates, and the normal incubation period is 29–30 days, or slightly shorter than in the other northern swans (Evans, 1975). Although a fledging period of only 30–45 days was once reported for this species (Dementiev and Gladkov, 1967), this is very doubtful in view of the nearly 70-day fledging period known to occur in the similarly tundra-breeding and conspecific whistling swan. Captive-raised birds also fledge in 60–70 days, which is comparable to the typical fledging period of whistling swans. Nonetheless, since the breeding territories are abandoned in September, the cygnets can be little more than 80–90 days old before they must begin their thousand-mile southward migration (Dementiev & Gladkov, 1967) Observations on wintering birds at the Wildfowl Trust indicate that offspring will rejoin their parents until as long as their fifth winter of life, and will depart with them in the spring (Kear, 2005).

Some information on the Bewick's swan relative to annual survival can be obtained from the returns of individually recognized birds to the Wildfowl Trust in later years. Evans (1970) provided a listing of such sightings for a seven-year period for birds that were adults or second-year birds when first sighted and recognized individually. Of a total of 792 identifiable birds, 287 were seen the subsequent winter season, indicating a minimum annual survival rate of 36.2 percent. However, 33 percent

Bewick's (tundra) swan, adult wing-flapping

returned seven years after initially having been first identified. This astonishing number of birds at least nine years old indicates that the survival rate of these swans is quite high, in spite of local hunting during their long migration. Sightings of birds returning in the third and subsequent seasons suggest an annual survival rate of 87 percent, and a maximum known longevity of 27 years has been reported for captive birds (Kear, 2005).

Status. The population of Bewick's swans that breeds in eastern Russia and winters in northwestern Europe numbered in the early 1970s about 6,000–7,000 birds. Of these, about half wintered in the Netherlands, about 1,500 in England, 500–1,000 in Ireland, 700 in Denmark, and 300 in West Germany. By 1990 the western breeding population had increased to about 29,000 birds, while the eastern Asian population totaled some 86,000 (Rose and Scott, 1994; Wetland International 2002). The western wintering populations have been stable in recent years, and wintering in England has increased appreciably with local protection and management. Although the birds are protected everywhere, illegal shooting results in about one fourth of the live birds carrying lead shot in their bodies. The restricted winter quarters for this population make it susceptible to future reductions, and the extensive tundra breeding grounds of the species are perhaps not yet in immediate danger, although global warming effects in the arctic will be an increasing problem for all arctic birds.

Relationships. The Bewick's swan and whistling swan are currently (2016) classified as conspecific by most authorities, but this interpretation may be an oversimplification of the evolutionary history of this group of swans. The similar small size of these two forms may simply reflect evolutionary convergence to comparable arctic tundra environments from separate stocks, rather than prove conspecificity (Johnsgard, 1974a). For example, Livezey (1996) judged the Bewick's swan to be the whooper swan's nearest relative.

Suggested readings. Dementiev & Gladkov, 1967; Scott & the Wildfowl Trust, 1972; del Toyo, Elliott & Sargatal, 1992; Todd, 1996; Kear, 2005.

Coscoroba Swan *Coscoroba coscoroba* (Molina) 1782

Other vernacular names. None in general English use. Koskorobaschwan (German); cygne coscoroba (French); cisne coscoroba (Spanish).

Subspecies and range. No subspecies recognized. Breeds in temperate South America from about 45 degrees South Latitude southward to Cape Horn and occasionally to the Falkland Islands, where it has rarely bred (Woods, 1975). Winters variably northward, with some birds reaching central Chile, northern Argentina, Paraguay, Uruguay, and extreme southern Brazil.

Measurements and weights (mostly from Scott & the Wildfowl Trust, 1972). Folded wing: both sexes 427–80 mm. Culmen: both sexes 65–70 mm; females only, 63–68 mm. Weights: males, 3.8–5.4 kg (ave. 4.6 kg); females, 3.2–4.5 kg (ave. 3.8 kg). Eggs: ave. 91 x 63 mm, whitish cream, 185 g.

Identification and field marks. Length 35–45" (90–115 cm). Coscoroba swans are both swan-like and goose-like; the adult plumage is entirely white except for the distal portion of the primaries, which is black. The head is feathered in front of the eyes, as in geese, but the bill is duck-like in shape and bright pinkish red, while the feet and legs are fleshy pink. The iris is yellow to reddish in adult males. *Females* have a dark brown iris color, as do immatures. *Juvenile* birds have a distinctive brownish pattern on the head and upper parts, and their bill is grayish. Some gray or brownish feathers may persist, especially on the wings, until the second year of life.

In the field, coscoroba swans somewhat resemble large white domestic geese, but their black wingtips (often scarcely visible except in flight) and bright pinkish red bills are distinctive. Both sexes have a loud *cos-cor-ooo* trumpeting call, uttered in flight and when on water.

NATURAL HISTORY

Habitat and foods. The favored habitats of the coscoroba swan are the lagoons and swampy areas of Chile's southern (Magellanic) region, where they are often found with black-necked swans. Weller (1967) found them to be very common on a large fresh-water marsh with its shallow areas dominated by bulrushes and cutgrass *(Zizaniopsis)*. These birds normally feed by swimming or wading in shallow water, but rarely also come ashore to graze along the water's edge. These foraging traits, in conjunction with the duck-like bill structure, must effectively reduce competition with the black-necked swan for food. However, little is known specifically of their food; Johnson (1965) stated that they eat various plants, aquatic insects, fish spawn, and apparently sometimes also small crustaceans. Todd (1996) saw birds apparently feeding on marine invertebrates along the coast.

Social behavior. Coscoroba swans seem most often to be seen in small numbers; rarely are flocks numbering in the hundreds ever mentioned. Weller (1967) reported loafing groups of 60–100 along

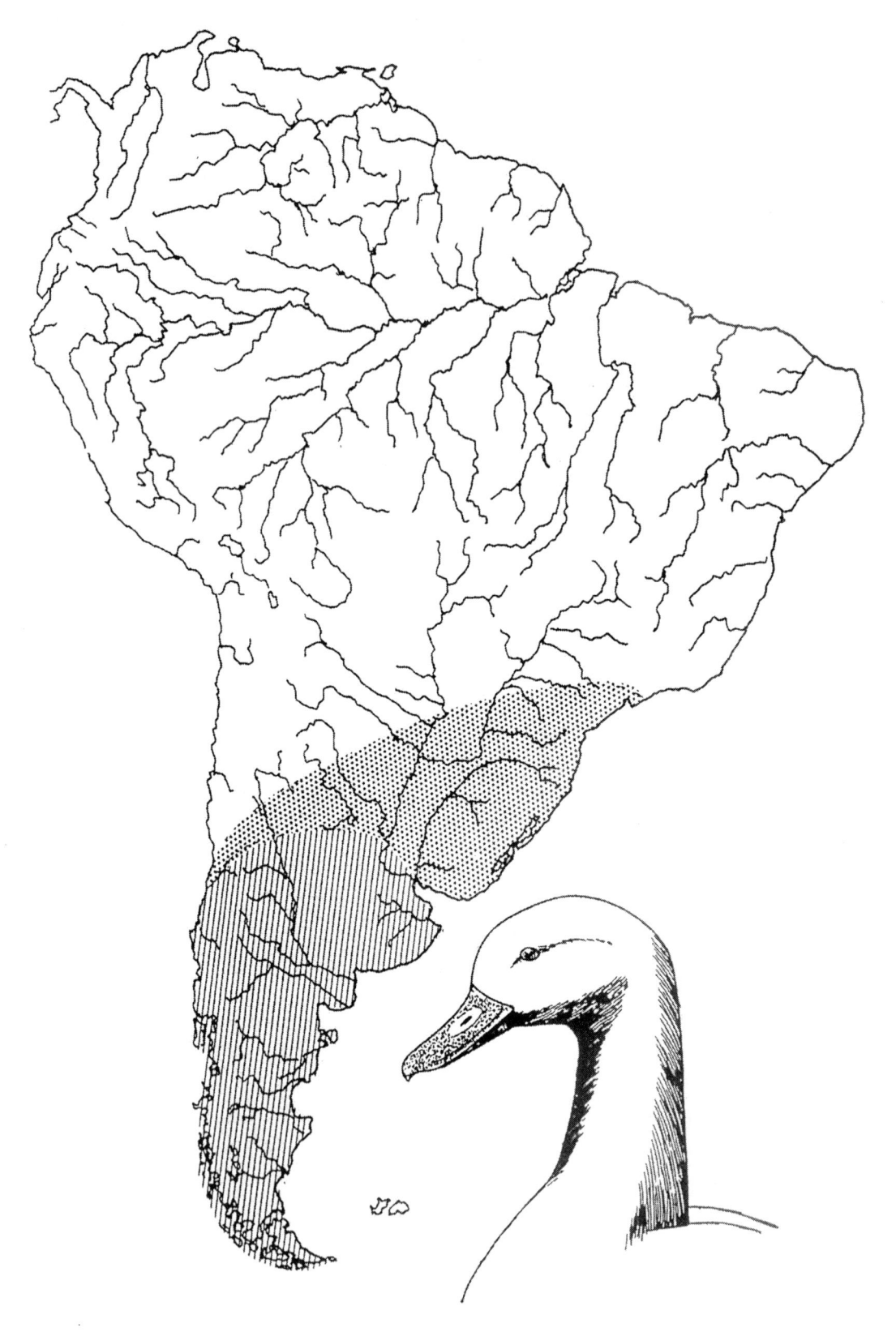

Map 9. Breeding or residential (hatched) and wintering (stippling) distributions of the coscoroba swan (from Johnsgard, 1978). Increasingly migratory northward in the southern parts of the range. Occasional to rare on the Falkland Islands.

Coscoroba swan, adult and cygnet

Coscoroba swan, adult pair

Coscoroba swan, head and bill detail

the shore of an Argentine lake; probably the species' greatest numbers are reached on various mid-summer molting grounds. It is presumed that these flocks consist primarily of paired birds, and that the pair bonds of this swan are permanent. Little is known definitely of sexual maturation rates; although captive birds have been known to breed at four years, it is assumed that the normal period to sexual maturity in the wild may be only two or perhaps at most three years. Unlike the typical swans, coscoroba swans seemingly do not have a triumph ceremony by which pair bonds are established and maintained. A simple greeting ceremony of calling does exist, and a threat display resembling that of the mute swan also occurs, but the means and timing of pair formation remain obscure. In further contrast to the other swans, copulation occurs while the birds stand in shallow water, after the male performs head-dipping. Afterwards both birds call as the male partially raises his folded wings. In this respect, and in lacking a triumph ceremony, the coscoroba swan strongly resembles some of the whistling ducks, such as the black-bellied (Johnsgard, 1965a).

Reproductive biology. The coscoroba swan nests in the austral spring; in Chile the breeding season extends from October to December (Johnson, 1965), whereas Gibson (1920) reported nesting in the

Fig. 14. Coscoroba swan, adults in flight

vicinity of Buenos Aires during the period of June to November. Although normally the bird is a solitary nester, he mentioned finding 16 nests in an area measuring 400 square yards. Johnson (1965) described the nest site as usually a bulky mound of soft vegetation placed among reed beds, in long grass close to water, or, if possible, on small islands. The nest cup is provided with an extensive down lining and, although only the female incubates the nest, the male closely guards it. The clutch size ranges from 4–7 eggs and probably averages about six. It has been reported that the female usually leaves her nest to forage for only about an hour in the morning and afternoon during incubation, and before leaving carefully covers the eggs with down or even twigs.

Incubation requires approximately 35 days, and all of the cygnets normally hatch on the same day. They differ from those of typical swans in being more distinctly patterned, with head markings slightly suggestive of those of whistling ducks. Adult coscoroba swans have never been seen carrying their young on their backs. They probably normally brood them on shore and thus back-carrying is less likely than in the black-necked swan. The fledging period in captivity is about 14 weeks, with wild birds flying when only 13 weeks old. The flightless period of the adults appears to be fairly variable; Weller (1975a) reported seemingly flightless birds on Tierra del Fuego in January or February, and elsewhere in Argentina they have been collected from mid-November to mid-April. The period to sexual maturity is uncertain, although one captive pair bred at three years of age, The average life expectancy in captivity is 7.3 years (Kear, 2006), but the maximum longevity record for captive birds is 20 years (Scott & the Wildfowl Trust, 1972).

Status. The coscoroba swan is not extremely abundant anywhere in South America, but neither is it obviously declining. As with the black-necked swans, loss of its favored temperate marsh habitats might be the most serious threat to its continued survival. It is not particularly sought-after by hunters, although some trampling of nests by cattle probably occurs locally. Up to 15,000 were once seen on a single marsh in the Mendoza Province of Argentina (Todd, 1996). It is apparently becoming more numerous on the Falkland Islands, where it is considered a rare visitor (Woods, 1975), and is known to have very rarely nested there. Recent population estimates suggest a stable world population of more than 25,000 birds (Rose and Scott, 1997; Wetlands International, 2002)

Relationships. It is impossible to neatly pigeonhole this species taxonomically, since it exhibits an unusual combination of goose, swan, and whistling duck characteristics (Johnsgard, 1965). Its skeletal anatomy suggests that it is a swan with some gooselike characteristics (Woolfenden, 1961), and except for the apparent lack of a triumph ceremony, its behavior is generally swan-like. A study of its feather proteins using electrophoresis data (Brush, 1976) indicated that the coscoroba swan should be retained within the tribe of geese and swans (Anserini), rather than being considered an evolutionary link with the whistling ducks (tribe Dendrocygnini). Livezey (1996) also concluded on morphological grounds that *Coscoroba* is a sister-group to *Cygnus*.

Suggested readings. Johnson, 1965; Weller, 1967; Scott & the Wildfowl Trust; 1972; del Toyo, Elliott & Sargatal, 1992; Todd, 1996; Kear, 2005.

III.

References

MONOGRAPHS ON SWANS AND OTHER ANATIDAE

Baldassarre, G. 2014. *Ducks, Geese and Swans of North America.* Revised & enlarged ed. Baltimore, MD: Johns Hopkins University Press. 2 vols.

Bauer, K. M., and U. N. Glutz von Blotztheim. 1968–1969. *Handbuch der Vogel Mitteleuropas.* Bands 2 & 3. Frankfurt am Main: Akademische Verlagsgesellschaft.

Bellrose, F. C. 1976. *Ducks, Geese and Swans of North America.* 2nd ed. Harrisburg, PA: Stackpole.

Bent. A. C. 1923. *Life Histories of North American Wild Fowl. Part 1.* U.S. National Museum Bulletin 126. Washington, DC: U.S. Government Printing Office.

Bent. A. C. 1925. *Life Histories of North American Wild Fowl.* Part 2. U.S. National Museum Bulletin 130. Washington, DC: U.S. Government Printing Office

Delacour, J. 1954-64. *The Waterfowl of the World.* 4 vols. London, UK: Country Life. (Swans are in vol. 1, 1954.)

Johnsgard, P. A. 1965. *Handbook of Waterfowl Behavior.* Ithaca, NY: Cornell University Press. 378 pp.

Johnsgard, P. A. 1968. *Waterfowl: Their Biology and Natural History.* Lincoln, NE: University of Nebraska Press. 138 pp.

Johnsgard, P. A. 1975. *Waterfowl of North America.* Bloomington, IN: Indiana University Press, 575 pp. http://digitalcommons.unl.edu/biosciwaterfowlna/1

Johnsgard, P. A. 1978. *Ducks, Geese and Swans of the World.* Lincoln, NE: University of Nebraska Press, Lincoln, NE. 404 pp. http://digitalcommons.unl.edu/biosciducksgeeseswans/

Johnsgard, P. A. 1987. *Waterfowl of North America: The Complete Ducks, Geese and Swans* (author of species descriptions only), with Robin Hill [artist], S. D. Ripley & The Duke of Edinburgh. Augusta, GA: Morris Publ. Co. 135 pp.

Kear, J. 2005. *Ducks Geese and Swans.* Oxford, UK: Oxford University Press. 2 vols. 910 pp.

Kortright, F. H. 1942. *The Ducks, Geese and Swans of North America.* Harrisburg, PA: Stackpole, and Washington, DC: Wildlife Management Institute.

Madge, S. C, and H. Burn. 1988. *Waterfowl: An Identification Guide to the Ducks, Geese and Swans of the World.* Boston, MA: Houghton Mifflin.

McCoy, J. J. 1967. *Swans.* New York, NY: Lothrop, Lee & Shepard.

Owen, M. 1977. *Wildfowl of Europe.* London, UK: Macmillan.

Palmer, R. S. (ed.). 1976. *Handbook of North American Birds.* Vols. 2 & 3. New Haven, CN: Yale University Press.

Price, A. L. 1995. *Swans of The World, In Nature, History, Myth and Art.* Tulsa, OK: Council Oak Distribution. 176 pp.

Scott, D. 1995. *Swans,* Minneapolis, MN: Voyageur Press. 72 pp.

Scott, P. (ed.), & The Wildfowl Trust. 1972. *The Swans.* London, UK: Michael Joseph.

Scott, P., & Boyd, H. 1957. *Wildfowl of the British Isles.* London, UK: Country Life.

Schull, M. 2012. *The Swan: A Natural History.* Ludlow, Shrop., UK: Merlin Unwin Books. 224 pp.

Todd, F. S. 1979. *Waterfowl: Ducks, Geese and Swans of the World.* New York, NY: Harcourt Brace Jovanovich.

Todd, F. S. 1996. *Natural History of the Waterfowl.* Vista, CA: Ibis Press. 490 pp.

Wilmore, S. B. 1974. *Swans of the World.* New York, NY: Taplinger.

OTHER MULTI-SPECIES REFERENCES

Atkinson-Willes, G. (ed.). 1963. *Wildfowl in Great Britain.* London, UK: Monographs of the Nature Conservancy, No.3.

American Ornithologists' Union. 1998. *Check-list of North American Birds.* 7th edition. American Ornithologists' Union, Washington, DC.

Bart, J., S. L. Earnst, and P. J. Bacon. 1991. Comparative demography of the swans: a review. *International Swan Symposium* 3:15–21.

Bottjer, P. D. 1983. Systematic relationships among the Anatidae: An immunological study, with a history of anatid classification, and a system of classification. Ph.D. Dissertation, Yale Univ., New Haven, CT.

Boyd, H. 1972. Classification. Pp. 17–28, in *The Swans,* P. Scott (ed.). Boston, MD: Houghton Mifflin.

Brush, A. H. 1976. Waterfowl feather proteins: Analysis of use in taxonomic studies. *Journal of Zoology* 179: 467-98.

Campbell, R. W., N. K. Dawe, I. McTaggart-Cowan, J. M. Cooper, G. W. Kaiser, and M. C. E. McNall. 1990. T*he Birds of British Columbia,* Volume 1: *Nonpasserines, Introduction and Loons through Waterfowl.* Vancouver, BC: Univ. British Columbia Press.

Clements, J. C.. 2000. *Birds of the World — A Checklist.* 5th ed. Vista, CA: Ibis Publishing Company.

Collins, D. P., C. A. Palmer, and R. E. Trost. 2011. *2011 Pacific Flyway Data Book.* Portland, OR: Division of Migratory Bird Management, U.S. Fish and Wildlife Service.

Cramp, S., and K. E. L. Simmons (chief eds.). 1977. *Handbook of the Birds of Europe, the Middle East, and North Africa,* Vol. 1. Oxford, U.K.: Oxford Univ. Press, 722 pp.

del Hoyo J., A. Elliott, and J. Sargatal. 1992. *Handbook of the Birds of the World. Vol.* 1. Barcelona, Spain: Lynx Editions, 696 pp.

Delacour, J., and E. Mayr. 1945. The family Anatidae. *Wilson Bull.* 57: 3-55.

Dementiev, G. P., & Gladkov, N. A. (eds.). 1967. *Birds of the Soviet Union.* Vol. 2. Washington, DC: Translated by Israel Program for Scientific Translations, U.S. Dept. Interior & Natl. Science Foundation.

Dickinson, E. (ed.) 2013. *The Howard and Moore Complete Checklist of the Birds of the World.* Eastbourne, East Sussex, UK: Aves Press.

Fay, F. H. 1960. The distribution of waterfowl to St. Lawrence Island, Alaska. *Wildfowl Trust Annual Report,* 12: 70-80.

Gabrielson, I. N., and F. C. Lincoln. (1959). *The Birds of Alaska.* Washington, D.C.: Wildlife Management Institute, and Harrisburg, PA: Stackpole.

Gibson, D. D., and G. V. Byrd. 2007. *Birds of the Aleutian Islands, Alaska.* Nuttall Ornithological Club, Cambridge, Massachusetts, and American Ornithologists' Union, Washington, DC.

Godfrey, W. E. 1986. *The Birds of Canada.* Revised ed. Ottawa, ON: National Museum of Natural Sciences.

Hanson, H. C., P. Queneau, and P. Scott. 1956. *The Geography, Birds and Mammals of the Perry River Region.* Montreal, QB: Arctic Institute of North America.

Hickey, J. J. 1952. *Survival Studies of Banded Birds.* U.S. Dept. of Interior, Fish and Wildlife Service, Special Scientific Report: Wildlife, No. 15. 177 pp.

Hudson, R. (ed.). 1975. *Threatened Birds of Europe.* London, UK: Macmillan.

Jacob, J., and A. Glaser. 1975. Chemotaxonomy of Anseriformes. *Biochem. Syst. Ecol.* 2: 215-220.

Johnsgard, P. A. 1960. Hybridization in the Anatidae and its taxonomic implications. *Condor* 62: 25-33. http://digitalcommons.unl.edu/biosciornithology/71

Johnsgard, P. A. 1961a. The taxonomy of the Anatidae—A behavioural analysis. *Ibis* 103: 71-85. http://digitalcommons.unl.edu/johnsgard/29

Johnsgard, P. A. 1961b. Tracheal anatomy of the Anatidae and its taxonomic significance. *Wildfowl* 12: 58-69.

Johnsgard, P. A. 1962. Evolutionary trends in the behaviour and morphology of the Anatidae. *Wildfowl* 13:130-148.

Johnsgard, P. A. 1963. Behavioral isolating mechanisms in the family Anatidae. *Proc. XIIIth International Ornithological Congress,* pp. 531-543. http://digitalcommons.unl.edu/johnsgard/23

Johnsgard, P. A. 1972. Observations on sound production of the Anatidae. *Wildfowl* 22:46-59. http://digitalcommons.unl.edu/johnsgard/13

Johnsgard, P. A. 1974. The taxonomy and relationships of the northern swans. *Wildfowl* 25: 155-61. http://digitalcommons.unl.edu/johnsgard/11

Johnsgard, P. A. 1979. *Anseriformes* (Anatidae and Anhimidae), Pp. 425-506, in *Check-list of The Birds of the World* (E. Mayr, ed.). 2nd ed. Cambridge, MA: Harvard Univ. Press. http://digitalcommons.unl.edu/johnsgard/32

Johnsgard, P. A. 2012a. *Wetland Birds of the Central Plains: South Dakota, Nebraska and Kansas.* Zea Books & Univ. of Nebraska Digital Commons, Lincoln, NE: http://digitalcommons.unl.edu/zeabook/8. 276 pp. Print edition available from http://www.lulu.com

Johnsgard, P. A. 2012b. *Wings over the Great Plains: Bird Migrations in the Central Flyway.* Zea E-Books & Univ. of Nebraska Digital Commons, Lincoln, NE:. http://digitalcommons.unl.edu/zeabook/13. 245 pp. Print edition available from http://www.lulu.com

Johnsgard, P. A. 2013. The swans of Nebraska. *Prairie Fire*, January 2013, pp. 12-13. http://www.prairiefirenewspaper.com/2013/01/the-swans-of-nebraska

Johnsgard, P. A., and J. Kear. 1968. A review of parental carrying of young by waterfowl. *Living Bird* 7:89-102.

Kessel, B., and D. G. Gibson. 1978. *Status and Distribution of Alaska Birds.* Studies in Avian Biology 1. Los Angeles, CA: Cooper Ornithological Society.

Livezey, B. C. 1986. A phylogenetic analysis of Recent anseriform genera using morphological characters. *Auk* 103: 737-754.

Livezey, B. C.1996. A phylogenetic analysis of the geese and swans (Anseriformes, Anserinae). *Systematic Biology* 45(4): 415–450.

Lutmerding, J. A., and A. S. Love. 2011. *Longevity Records of North American Birds.* Version 2011.2. Laurel, MD: Bird Banding Laboratory, Patuxent Wildlife Research Center.

Madsen, C. S., K. P. McHough, and S. R. De Kloet. 1988. A partial classification of waterfowl (Anatidae) based on single-copy DNA. *Auk* 105: 452-459.

Marchant, S., and P. J. Higgins (coordinators). 1990. *Handbook of Australian, New Zealand and Antarctic Birds.* Volume 1, part B. Melbourne, Victoria: Oxford Univ. Press.

Martin, A, H. S. Zim and A.L. Nelson. 2011. *Wildlife and Their Food Plants.* Reprint of 1951 ed. New York, NY: Dover.

Meyer de Schauensee, R, 1966, *The Species of Birds of South America, with their Distribution.* Philadelphia, PA: Academy of Natural Sciences, and Narberth, PA: Livingston Pub.

Murie, O. J. 1959. *Fauna of the Aleutian Islands and Alaska Peninsula.* Washington, DC: U.S. Dept. of Interior, Fish and Wildlife Service, North American Fauna, No. 61: 1-406.

Ogilvie, M. A. 1972. Distribution, numbers and migration. Pp. 29-56, in *The Swans* (P. Scott, ed.). Boston, MA: Houghton Mifflin.

Petzold, H.-G. 1964. Beiträge zur vergleichenden Ethologie der Schwane (Anseres, Anserini). *Beiträge zur Vogelkunde* 10: 1-123.

Raftovich, R. V., K. A. Wilkins, K. D. Richkus, S. S. Williams, and H. L. Spriggs. 2010. *Migratory Bird Hunting Activity and Harvest during the 2008 and 2009 Hunting Seasons.* Laurel, MD: U.S. Fish and Wildlife Service.

Rees, E C., P. Lievesley, R. A. Pettifor, and C. Perrins. 1996. Mate fidelity in swans: an interspecific comparison. Pp. 118–137, in *Partnerships in Birds: the Study of Monogamy.* London, UK: Oxford University Press.

Rose, P. M., and D. A. Scott. 1997. *Waterfowl Population Estimates.* 2nd edition. Wetlands International Publication 44. Wetlands International, Wageningen, Netherlands.

Sauer, J. R., J. E. Hines, J. E. Fallon, K. L. Pardieck, D. J. D. J. Ziolkowski, and W. A. Link. 2014. *The North American Breeding Bird Survey, Results and Analysis 1966–2013. Version 01.30.2015* Laurel, MD: USGS Patuxent Wildlife Research Center. http://www.mbr-pwrc.usgs.gov/bbs/

Sibley, C. G., and B. L. Monroe, Jr. 1990. *Distribution and Taxonomy of Birds of the World.* New Haven, CT: Yale Univ. Press.

Sibley, C. G., and J. E. Ahlquist. 1990. *Phylogeny and Classification of Birds: A Study in Molecular Evolution.* New Haven, CT: Yale Univ. Press.

Stejneger, L. 1882. Outline of a monograph of the Cygninae. *Proc. U.S. Nat. Mus.* 5:174–221.

U.S. Fish and Wildlife Service. 2007. *Productivity Surveys of Geese, Swans and Brant Wintering in North America, 2007.* Arlington, Virginia: Division of Migratory Bird Management, U.S. Fish and Wildlife Service.

Voous, K. H. 1960. *Atlas of European Birds.* London, UK: Elsevier, Nelson.

Weller, M. W. 1964a. General habits. Pp. 15-34, in *The Waterfowl of the World,* Volume 4 (J. Delacour, ed.). London, UK: Country Life.

Weller, M. W. 1964b. The reproductive cycle. Pp. 35-79, in *The Waterfowl of the World,* Volume 4 (J. Delacour, ed.). London, UK: Country Life.

Weller, M. W. 1964c. Ecology. Pp. 80-107, in *The Waterfowl of the World,* Volume 4 (J. Delacour, ed.). London, UK: Country Life.

Weller, M. W. 1964d. Distribution and species relationships. Pp. 108-120, in *The Waterfowl of the World,* Volume 4 (J. Delacour, ed.). London, UK: Country Life.

Weller, M. W. 1967. Notes on some marsh birds of Cape San Antonio, Argentina. *Ibis* 109: 391-441.

Weller, M. W. 1968. Notes on some Argentine anatids. *Wilson Bulletin* 80:189–212.

Weller, M. W. 1975. Habitat selection by waterfowl of Argentina. *Wilson Bull.* 87:83–90.

Weller, M. W. 1980. *The Island Waterfowl.* Ames, IA: Iowa State Univ. Press.

Wetlands International. 2002. *Waterfowl Population Estimates–Third Edition.* Wageningen, The Netherlands: Wetlands International, Global Series No. 12.

Wetlands International. 2006. *Waterbird Population Estimates.* 4th. ed. Wageningen, Netherlands: Wetlands International,

Wetmore, A. 1951. Observations on the genera of the swans. *J. Wash. Acad. Sci.* 41:338-340.

Woolfenden, G. E. 1961. Postcranial osteology of the waterfowl. *Bulletin of the Florida State Museum, Biological Sciences* 6:1-29.

INDIVIDUAL SPECIES REFERENCES

MUTE SWAN

Bacon, P. J. 1980. Status and dynamics of a mute swan population near Oxford between 1976 and 1978. *Wildfowl* 31:37–50.

Bannerman, D. A. 1957. *The Birds of the British Isles* Vol. 6. London, UK: Oliver & Boyd.

Bacon, P. J., and A. E. Coleman. 1986. An analysis of weight changes in the mute swan *Cygnus olor. Bird Study* 33:145–158.

Boase, H. 1959. Notes on the display, nesting and moult of the mute swan. *British Birds* 52:114-121.

Breault, A. 2004. Status and management of mute swans in southwest British Columbia (Abstract). 19th Trumpeter Swan Society Conference: 203.

Carey, C. G. 2000. Mute swan control and trumpeter swan experimental breeding project in urban central Oregon. 17th Trumpeter Swan Society Conference: 114-116.

Chesapeake Bay Mute Swan Working Group. 2004. *Mute Swan (Cygnus olor) in the Chesapeake Bay: A Baywide Management Plan.* Annapolis, MD: Maryland Department of Natural Resources.

Ciaranca, M. A., C. C. Allin, and G. S. Jones. 1997. Mute swan (*Cygnus olor*). *The Birds of North America 273.* The Birds of North America, Inc. Philadelphia, PA: The Academy of Natural Sciences, and Washington, DC: American Ornithologists' Union, 28 pp.

Coleman, A. E., and C. D. T. Minton. 1979. Pairing and breeding of mute swans in relation to natal area. *Wildfowl* 30:27–30.

Coleman, A. E., J. T. Coleman, P. A. Coleman, and C. D. T. Minton. 2001. A 39-year study of a mute swan *Cygnus olor* population in the English Midlands. *Ardea* 89 (Special Issue): 123–133.

Coleman, J. T., A. E. Coleman, and D. Elphick. 1994. Incestuous breeding in the mute swan *Cygnus olor. Ringing and Migration* 15:127–128.

Froelich, A. J., J. C. Johnson, and D. M. Lodge. 1999. Food preferences of mute and trumpeter swans. 16th Trumpeter Swan Society Conference: 133.

Gelston, W. L., and R. D. Wood. 1982. *The Mute Swan in Northern Michigan.* Traverse City, MI: Myers Print Service.

Harrison, J. G., and M. A. Ogilvie. 1968. Immigrant mute swans in south-east England. *Wildfowl Trust Annual Report* 18:85-87.

Hindman, L. J., R. A Malecki, and C. M. Sousa. 2004. Mute Swans in Maryland: their status and a proposal for management (Abstract). 19th Trumpeter Swan Society Conference: 204.

Hilprecht. A. 1956. *Höcherschwan, Singshwan, Zwergschwan.* Wittenberg: Neue Brehm Bücherei.

Huxley, J. S. 1947. Display of the mute swan. *Brit. Birds* 40:140-134.

Jobes, C. R. 1986. Energetics of growth of trumpeter and mute swan cygnets (Abstract). 9th Trumpeter Swan Society Conference: 119.

Johnson, W. C. (Joe). 1999. Observations of territorial conflict between trumpeter swans and mute swans in Michigan. 16th Trumpeter Swan Society Conference: 134-136.

Johnston, W. 1935. Notes on the nesting of captive mute swans. *Wilson Bulletin,* 47: 237-238.

Lumsden, H. G. 1986. The trumpeter swan/mute swan experiment: Ontario. 9th Trumpeter Swan Society Conference: 117-118.

Mathiasson, S. 1973. A moulting population of non-breeding mute swans with special reference to flight-feather moult, feeding ecology and habitat selection. *Wildfowl* 24:43–53.

Mathiasson, S. 1981. Weight and growth rates of morphological characters of *Cygnus olor*. *International Swan Symposium* 2: 379–389.

McCleery, R. H., C. Perrins, D. Wheeler, and S. Groves. 2002. Population structure, survival rates and productivity of mute swans breeding in a colony at Abbotsbury, Dorset, England. *Waterbirds* 25 (Special Publication 1):192–201.

Minton, C. D. T. 1968. Pairing and breeding of mute swans. *Wildfowl 19:* 41-60.

Munro, R. E., L. T. Smith, and J. J. Kupa. 1968. The genetic basis of color differences observed in the mute swan (*Cygnus olor*). *Auk* 85:504–505.

Nelson, H. K. 1999. Mute swan populations, distribution, and management issues in the United States and Canada. 16th Trumpeter Swan Society Conference: 125-132.

O'Brien, M., and R. A. Askins. 1985. The effects of mute swans on native waterfowl. *Connecticut Warbler* 5:27–31.

Ogilvie, M. A. 1967. Population changes and mortality of the mute swan in Britain. *Wildfowl Trust Annual Report* 18: 64–73.

Owen, M., and C. J. Cadbury. 1975. The ecology and mortality of mute swans at the Ouse Washes, England. *Wildfowl* 25:31–42.

Perrins, C. M., and C. M. Reynolds. 1967. A preliminary study of the mute swan, *Cygnus olor*. *Wildfowl Trust Annual Report,* 18:74-84.

Perrins, C. M., G. Cousquer, and J. Waine. 2003. A survey of blood lead levels in mute swans *Cygnus olor*. *Avian Pathology* 32:205–212.

Perry, M. C. (ed.) 2004. *Mute Swans and Their Chesapeake Bay Habitats: Proceedings of a Symposium.* Information and Technology Report USGS/BRD/ITR-2004-0005. Reston, VA: Biological Resources Disciplines, U.S. Geological Survey.

Perry, M. C., P. C. Osenton, and E. J. R. Lohnes. 2004. Food habits of mute swans in the Chesapeake Bay. Pp. 31–35, in Perry (2004).

Petrie, S. A. 2004. Review of the status of mute swans on the Canadian side of the Lower Great Lakes (Abstract). 19th Trumpeter Swan Society Conference: 201- 202.

Petrie, S. A. 2004. Review of the status of mute swans on the Canadian side of the lower Great Lakes. Pp. 23–27, in Perry (2004).

Petrie, S. A., and C. M. Francis. 2003. Rapid increase in the lower Great Lakes population of feral mute swans: a review and a recommendation. *Wildlife Society Bulletin* 31:407–416.

Rees, E C., P. Lievesley, R. A. Pettifor, and C. Perrins. 1996. Mate fidelity in swans: an interspecific comparison. Pp. 118–137, in *Partnerships in Birds: the Study of Monogamy*. London, UK: Oxford University Press.

Reese, J. G. 1975. Productivity and management of feral mute swans in Chesapeake Bay. *Journal of Wildlife Management* 39:280–286.

Reese, J. G. 1980. Demography of European mute swans in Chesapeake Bay. *Auk* 97:449–464.

Reichholf, V. J. 1984. On the function of territoriality in the mute swan (*Cygnus olor*). *Verhandlungen Ornithologischen Gesellschaft in Bayern* 24:125–135 [in German, with English summary].

Reiswig, B. 1986. Western mute swan population status and agency attitudes. 9th Trumpeter Swan Society Conference: 116.

Reynolds, C. M. 1965. The survival of mute swan cygnets. *Bird Study* 12:128–129.

Scott, D. K. 1984. Winter territoriality of mute swans *Cygnus olor*. *Ibis* 126:168–176.

Scott, D. K., and M. E. Birkhead. 1983. Resources and reproductive performance in mute swans *Cygnus olor*. *Journal of Zoology* 200:539–547.

Sears, J. 1989. Feeding activity and body condition of mute swans *Cygnus olor* in rural and urban areas of a lowland river system. *Wildfowl* 40:88–98.

Sousa, C. M., R. A. Malecki, A. J. Lembo, Jr., and L. J. Hindman. 2008. Monitoring habitat use by male mute swans in the Chesapeake Bay. *Proceedings of the Southeastern Association of Fish and Wildlife Agencies* 62:88–93.

Stone, W. B., and A. D. Masters. 1971. Aggression among captive mute swans. *New York Fish and Game Journal*, 17:50-52.

Tatu, K. S., J. T. Anderson, and L. J. Hindman. 2007. Predictive modeling for submerged aquatic vegetation (SAV) decline due to mute swans in the Chesapeake Bay. 20th Trumpeter Swan Society Conference: 140-147.

Tatu, K. S., J. T. Anderson, L. J. Hindman, and G. Seidel. 2007a. Diurnal foraging activities of mute swans in Chesapeake Bay, Maryland. *Waterbirds* 30:121–128.

Tatu, K. S., J. T. Anderson, L. J. Hindman, and G. Seidel. 2007b. Mute swans' impact on submerged aquatic vegetation in Chesapeake Bay. *Journal of Wildlife Management* 71:1431–1439.

Therres, G. D., and D. F. Brinker. 2004. Mute swan interactions with other birds in the Chesapeake Bay. Pp. 43–46, in Perry (2004).

Ticehurst, N. F. 1967. *The Mute Swan in England*. London, UK: Cleaver-Hume Press.

Willey, C. H. 1968. The ecology, distribution, and abundance of the mute swan (*Cygnus olor*) in Rhode Island. M.S. thesis, University of Rhode Island, Kingston, RI.

Willey, C. H., and B. F. Halla. 1972. *Mute Swans of Rhode Island*. West Kingston, RI: Division of Fish and Wildlife, Rhode Island Department of Natural Resources, Wildlife Pamphlet 8.

Wood, R., and W. L. Gelston. 1972. *Preliminary Report: The Mute Swans of Michigan's Grand Traverse Bay Region*. Report 2683. Lansing, MI: Wildlife Division, Michigan Department of Natural Resources.

BLACK SWAN

Braithwaite, L. W., and H. J. Frith. 1969. Waterfowl in an inland swamp in New South Wales. III. Breeding. *CSIRO Wildl. Res.* 14:1-36.

Frith, H. J. 1982. *Waterfowl in Australia*. 2nd ed. Honolulu, HI: East-West Center Press.

Guiler, E. R. 1966. The breeding of the black swan *Cygnus atratus* Latham in Tasmania, with special reference to its management problems. *Proc. Royal Soc. of Tasmania* 100:31–52.

Guiler, E. R. 1970. The use of breeding sites of black swan in Tasmania. *Emu* 70:3–8.

Hadden, D. 2010. *Waterbirds of Australia*. Sydney, NSW, Australia: New Holland. 96 pp.

Lavery, H. J. 1964. *An investigation of the biology and ecology of waterfowl in North Queensland*. M..Sc. thesis, University of Queensland, Brisbane, Australia.

Lavery, H. J. 1967. The black swan in Queensland. Brisbane, Australia: *Dept. Primary Industries, Div. of Plant Industries, Advisory Leaflet 911*.

Lavery, H. J. 1970. The comparative ecology of waterfowl in North Queensland. *Wildfowl* 21: 67–77.

Long, J. L. 1981. *Introduced Birds of the World.* Newton Abbott, Devon, UK: David & Charles. 58 pp.
Marchant, S., and P. J. Higgins (Coordinators). 1990. *Handbook of Australian, New Zealand & Antarctic Birds.* Vol. 1, Part B. Ratites to Ducks. Melbourne, Australia: Oxford University Press. 1,400 pp.
Oliver, W. R. B. 1930. *New Zealand Birds.* Wellington, NZ: Fine Arts.
Pringle, J. D. 1985. *The Waterbirds of Australia.* Sidney, Australia: Angus and Robertson.

BLACK-NECKED SWAN

Blake, E. R.1977. *Manual of Neotropical Birds.* Vol. 1. Chicago, IL: University of Chicago Press. 674 pp.
Cawkell, E. M., & Hamilton, J. E. 1961. The birds of the Falkland Islands. *Ibis* 103a: 1–27.
Cobb, A. F. 1933. *Birds of the Falkland Islands.* London, UK: Witherby.
Crayshaw, B. 1907. *The Birds of Tierra del Fuego.* London, UK: Bernard Quaritch.
Johnson, A. W. 1965. *The Birds of Chile, and Adjacent Regions of Argentina, Bolivia and Peru.* Vol. 1. Buenos Aires. Argentina: Platt Establecementios Graficos. 397 pp.
Rossi, J. A. H. 1953. Contribución al conocimiento de la biologia del cisne de cuello negro. *El Hornero* 10:1–17.
Schlatter, R.P., J. Salazar, A. Villa, and J. Meza. 1991. Reproductive biology of the black-necked swan, *Cygnus melanocoryphus*, at three Chilean wetland areas, and feeding ecology at Rio Cruces. *Wildfowl* Spec. Suppl. 1: 88–94.
Weller, M. W. 1967. Notes on some marsh birds of Cape San Antonio, Argentina. *Ibis* 109:391-416.
Weller, M. W. 1975, Habitat selection by waterfowl of Argentine Isla Grande. *Wilson Bulletin* 87:83-90.
Williams, M. J. 1973. Mortality of the black swan in New Zealand—a progress report. *Wildfowl* 24:94–102
Williams, M. J. 1981. The demography of New Zealand's black swan *Cygnus atratus* population. Pp. 147–156, in *Second IWRB Swan Symposium*, Sapporo, Japan. 1980. Slimbridge, UK: International Waterfowl Research Bureau [IWRB].

TRUMPETER SWAN

By way of bibliographic explanation, The Trumpeter Swan Society (TTSS) is a North American non-profit organization formed in 1968 that is devoted to the conservation and restoration of trumpeter swans. The proceedings of its periodic conferences cited here are deposited in the Society's office in Plymouth, Minnesota (763-694-1248; email: ttss@trumpeterswansociety/org). The proceedings are housed with the Three Rivers Park District, 12615 Rockford Road, Plymouth, MN 55441-1248, and can be ordered from the Society (www.trumpeterswansociety.org). From 1971 through 1996 the Society published a *Newsletter* for the members. In 1991 another publication, *Trumpetings*, was produced that became the member newsletter in 1997. Also in 1997, TTSS initiated *North American Swans: Bulletin of The Trumpeter Swan Society.* Starting in 1997, conference proceedings have been published as issues of *North American Swans.*

Anderson, P. S. 1992. Changing land use and trumpeter swans in the Skagit Valley. 13th Trumpeter Swan Society Conference: 150-156.
Anderson, P. S. 1993. Distribution and habitat selection by wintering trumpeter swans *Cygnus buccinator* in the Skagit Valley, Washington. M.S. thesis, University of Washington, Seattle, WA.
Anderson, P. S. 1994. Distribution and habitat selection by wintering trumpeter swans in the lower Skagit Valley, Washington. 14th Trumpeter Swan Society Conference: 61-71.
Anderson, P. S. 2004. The Pacific Coast Population – historical perspective and future concerns. 19th Trumpeter Swan Society Conference: 3-8.
Andrews, R., and D. Hoffman. 2007. Iowa's trumpeter swan restoration program – a 2005 update. 20th Trumpeter Swan Society Conference: 3-10.
Anglin, R. M. 1999. Rocky Mountain Population of trumpeter swans: The winter range expansion program. 16th Trumpeter Swan Society Conference: 56-58.

Babineau, F. M. 2004. Winter ecology of trumpeter swans in southern Illinois. Master's thesis, Southern Illinois University, Carbondale, IL.

Babineau, F. M., and D. Holm. 2004. Winter distribution and habitat use of trumpeter swans in Illinois. 19th Trumpeter Swan Society Conference: 166-174.

Bailey, T. N., E. E. Bangs, and M. F. Portner. 1986. Trumpeter swan surveys and studies on the Kenai National Wildlife Refuge and Kenai Peninsula, Alaska, 1957–1984. (Abstract). 9th Trumpeter Swan Society Conference: 64.

Bailey, T. N., M. F. Portner, E. E. Bangs, W. W. Larned, R. A. Richey, and R. L. Delaney. 1990. Summer and migratory movements of trumpeter swans using the Kenai National Wildlife Refuge, Alaska. 11th Trumpeter Swan Society Conference: 72-81.

Bales, B. 1992. Update on the Pacific Coast Population Swan Management Plan. 13th Trumpeter Swan Society Conference: 131-132.

Bales, B. D., and D. Kraege. 1992. Management challenges related to Pacific Coast Population trumpeter swans in Oregon and Washington. 13th Trumpeter Swan Society Conference: 157-159.

Bales, B. D., and D. Kraege. 1994. Management challenges in the 1990's related to Pacific Coast Population trumpeter swans in Oregon and Washington. 14th Trumpeter Swan Society Conference: 98-100.

Ball, I. J., E. O. Garton, and R. E. Shea. 2001. History, ecology and management of the Rocky Mountain Population of trumpeter swans: Implications for restoration. 17th Trumpeter Swan Society Conference: 45-49.

Banko, W. 1960. *The Trumpeter Swan: Its History, Habits, and Population in the United States.* Washington, D.C.: U.S. Dept. of Interior, Fish and Wildlife Service, North American Fauna No. 63. 214 pp.

Barrett, V. A., and E. R. Vyse. 1982. Comparative genetics of three trumpeter swan populations. *Auk* 99:103–108.

Bartok, N. D. 2004. 2002 nesting success of the trumpeter swan (*Cygnus buccinator*) population that frequents the Wye Marsh, Ontario. 19th Trumpeter Swan Society Conference: 159-165.

Bauer, R. D. 1991. U. S. Fish and Wildlife Service involvement in winter management of the Rocky Mountain Population of trumpeter swans. 12th Trumpeter Swan Society Conference: 195.

Becker, D. and J. S. Lichtenberg. 2004. Trumpeter swan reintroduction on them Flathead Indian Reservation – an overview and update. 19th Trumpeter Swan Society Conference: 128-133.

Becker, D. and J. S. Lichtenberg. 2007. Trumpeter swan reintroduction on the Flathead Indian Reservation. 20th Trumpeter Swan Society Conference: 100-105.

Becker, D. M. 2001. Trumpeter swan reintroduction on the Flathead Indian Reservation 17th Trumpeter Swan Society Conference: 103-106.

Berquist, J. 1990. Status report on Turnbull National Wildlife Refuge's trumpeter swan population. 11th Trumpeter Swan Society Conference: 117.

Beyersbergen, G. W., M. Heckbert, R. Kaye, T. Sallows, and P. Latour. 2007. The 2005 international survey of trumpeter swans in Alberta, Saskatchewan, Manitoba, and the Northwest Territories. 20th Trumpeter Swan Society Conference: 78-87.

Beyersbergen, G. W., and L. Shandruk. 1994. Interior Canada Subpopulation of trumpeter swans – status 1992. 14th Trumpeter Swan Society Conference: 103-110.

Beyersbergen, G. W., and R. Kaye. 2007. Elk Island National Park trumpeter swan Reintroduction - 2005 update. 20th Trumpeter Swan Society Conference: 88-97.

Blus, L. J., R. Stroud, B. Reiswig, and T. McEneaney. 1990. Lead poisoning and other mortality factors of trumpeter swans (Abstract). 11th Trumpeter Swan Society Conference: 152.

Bogdan, L. 2004. Development of a detailed landscape plan to support overwintering and migrating waterfowl for the Fraser River delta (Abstract). 19th Trumpeter Swan Society Conference: 15.

Bouffard, S. H. 1986a. Rocky Mountain Population (Tristate flock): Status of trumpeter swans at Camas National Wildlife Refuge 1983-1984. 9th Trumpeter Swan Society Conference: 54- 55.

Bouffard, S. H. 1986b. Pacific Coast Population: Status of trumpeter swan restoration flocks, 1983-84. 9th Trumpeter Swan Society Conference: 89-91.

Bouffard, S. H. 2000. Recent changes in winter distribution of RMP [Rocky Mountain Population] trumpeter swans. 17th Trumpeter Swan Society Conference: 53-59.

Boyd, S. 1994. Abundance patterns of trumpeter swans and tundra swans on the Fraser River Delta, British Columbia (Abstract). 14th Trumpeter Swan Society Conference: 48.

Boyd, W. S., and A. Breault. 2004. Trumpeter swans wintering in southwest British Columbia: An assessment of status and trends (Abstract). 19th Trumpeter Swan Society Conference: 10.

Brown, C. S., and J. Luebbert. 2000. Field triage and rehabilitation of swans. 17th Trumpeter Swan Society Conference: 170-76.

Brown, S. 1990. A status report of the introduced trumpeter swan population at Ruby Lakes National Wildlife Refuge, Nevada. 11th Trumpeter Swan Society Conference: 123-124.

Buffet, D. and M. R. Petrie. 2004. Spatial and temporal use of estuary and upland habitat by waterfowl wintering on the Fraser River delta and North Puget Sound (Abstract). 19th Trumpeter Swan Society Conference: 15.

Burgess, H. H. 1986. Potential trumpeter swan restoration. 9th Trumpeter Swan Society Conference: 97-111.

Burgess, H. H. 2001a. History of the High Plains trumpeter swan restoration. *North American Swans* 30(1): 6-14.

Burgess, H. H. 2001b. North Dakota trumpeter swan observations. *North American Swans* 30(1): 21-24.

Burgess, H. H. 2002a. Trumpeter swan myths, movements and migrations of the High Plains Flock. *North American Swans* 31:7–11.

Burgess, H. H. 2002b. High Plains trumpeter swan nesting ecology. *North American Swans* 31(1): 5-6.

Burgess, H. H., and M. E. Bote. 1999. Observations of trumpeter swans in Manitoba. *North American Swans* 28(1): 25-30.

Burgess, H. H., and R. Burgess. 1988. Elk Island National Park trumpeter swan restoration experimental project. 10th Trumpeter Swan Society Conference: 78-88.

Burgess, H. H., and R. Burgess. 1991. History of trumpeter swan restoration to the Upper Midwest. 12th Trumpeter Swan Society Conference: 131-132.

Burgess, H. H., and R. Burgess. 1997. Trumpeter swans once wintered in Texas – why not now? *North American Swans* 26(2): 50-53.

Burgess, H. H., and R. Burgess. 1998. The Nebraska trumpeter swans. *North American Swans* 27(1): 30-31.

Burgess, H. H., R. Burgess, and M. Bote. 1999a. Trumpeter swans once wintered on the lower Mississippi River. Why not now? 16th Trumpeter Swan Society Conference: 3-5.

Burgess, H. H., R. Burgess, and M. Bote. 1999b. Developing trumpeter swan wintering areas. 16th Trumpeter Swan Society Conference: 25-26.

Burgess, H. H., R. Burgess, and D. K. Weaver. 1990. Potential trumpeter swan restoration and expansion. 11th Trumpeter Swan Society Conference: 62-64.

Canniff, R. S. 1986. Wintering trumpeter swans, Skagit Valley, Washington: update 19801984. 9th Trumpeter Swan Society Conference: 71-75.

Canniff, R. S. 1990. Trumpeter and tundra swan collar sightings in the Skagit Valley, 19771978 to 1987-1988. 11th Trumpeter Swan Society Conference: 125-141.

Carey, C. G. 2000. Mute swan control and trumpeter swan experimental breeding project in urban central Oregon. 17th Trumpeter Swan Society Conference: 114-116.

Carrick, W. H. 1991. Use of imprinted swans to establish a migratory population. 12th Trumpeter Swan Society Conference: 143.

Carrick, W. H. 1999. Induced migration using ultralite aircraft. 16th Trumpeter Swan Society Conference: 115-116.

Central Flyway Council. 1991. Position statement on tundra swan hunting in the Central Flyway relative to potential conflicts with trumpeter swan restoration. 12th Trumpeter Swan Society Conference: 73-74.

Childress, D. 1986. Trumpeter swan expansion in Montana. 9th Trumpeter Swan Society Conference: 47.

Churchill, B. P. 1988. Potential trumpeter swan nesting habitat in northeastern British Columbia. 10th Trumpeter Swan Society Conference: 29-35.

Cochran, D. K. 1970. Food preferences in captivity and weight gains in trumpeter swans. *Trumpeter Swan Society Newsletter* 4:9–13.

Comeau-Kingfisher, S. and T. Koerner. 2007. Status of the High Plains flock of trumpeter swans in 2005. 20th Trumpeter Swan Society Conference: 23-27.

Compton, D. 1988. 1985. Captive trumpeter swan survey results. 10th Trumpeter Swan Society Conference: 146-148.

Compton, D. 1991a. Results of the 1988 captive trumpeter swan survey. 12th Trumpeter Swan Society Conference: 45.

Compton, D. 1991b. Trumpeter swan banding protocol – a survey of the banders. 12th Trumpeter Swan Society Conference: 49-52.

Compton, D. 1991c. Hennepin Parks trumpeter swan restoration update. 12th Trumpeter Swan Society Conference: 91-94.

Compton, D. 1991d. Transport of trumpeter swan eggs and cygnets. 12th Trumpeter Swan Society Conference: 147.

Compton, D. C. 1996. Interior Population status report, highlights, and trends, December 1994. 15th Trumpeter Swan Society Conference: 18-37.

Conant, B. 1991. Alaskan trumpeter swan status report. (Abstract). 12th Trumpeter Swan Society Conference: 9.

Conant, B., J. I. Hodges, D. J. Groves, and J. G. King. 1992. The 1990 census of trumpeter swans on Alaskan nesting habitats. 13th Trumpeter Swan Society Conference: 133-146.

Conant, B., J. I. Hodges, D. J. Groves, and J. G. King. 1994. A potential summer population of trumpeter swans (*Cygnus buccinator*) for Alaska. (Abstract). 14th Trumpeter Swan Society Conference: 5-6.

Conant, B., J. I. Hodges, D. J. Groves, and J. G. King. 1999. The 1995 census of trumpeter swans on Alaskan nesting habitats. 16th Trumpeter Swan Society Conference: 75-97.

Conant, B., J. I. Hodges, D. J. Groves, and J. G. King. 2000. The 1995 census of trumpeter swans on Alaskan nesting habitats (Abstract). 17th Trumpeter Swan Society Conference: 3.

Conant, B., J. I. Hodges, D. J. Groves, and J. G. King. 2007. The 2005 census of trumpeter swans on Alaskan nesting habitats (Abstract). 20th Trumpeter Swan Society Conference: 107-112.

Conant, B., J. I. Hodges, J. G. King, and S. L. Cain. 1988. Alaska trumpeter swan status report – 1985. 10th Trumpeter Swan Society Conference: 121-129.

Conant, B., J. I. Hodges, R. J. King, and A. Loranger. 1986. Alaska trumpeter swan status report – 1984. 9th Trumpeter Swan Society Conference: 76-89.

Cooper, B. A., and R. J. Ritchie. 1990. Migration of trumpeter and tundra swans in east-central Alaska during spring and fall, 1987. 11th Trumpeter Swan Society Conference: 82-91.

Cooper, B., J. King, and R. J. Ritchie. 1991. Swan migration routes in the Nelchina Basin, Alaska, during spring migration 1989 (Abstract). 12th Trumpeter Swan Society Conference: 11.

Cooper, J. A., and D. K. Weaver (eds.). 1986. *Trumpeter Swan Bibliography*. Available from the Trumpeter Swan Society, 12615 Rockford Road, Plymouth, MN 55441

Corace, R. G., III, D. L. McCormick, and V. Cavalieri. 2006. Population growth parameters of a reintroduced trumpeter swan flock, Seney National Wildlife Refuge, Michigan, USA (1991–2004*). Waterbirds* 29:38–42.

Cornely, J. E., S. P. Thompson, E. McLaury, and L. D. Napier. 1985. A summary of trumpeter swan production at Malheur National Wildlife Refuge, Oregon. *Murrelet* 66: 50–55.

Czarnowski, K. 1986. Yellowstone National Park policy for managing trumpeter swans. 9th Trumpeter Swan Society Conference: 27-28.

Degernes, L. A. 1991. The Minnesota trumpeter swan lead poisoning crisis of 1988-89. 12th Trumpeter Swan Society Conference: 114-118.

Degernes, L. A., and P. T. Redig. 1990b. Diagnosis and treatment of lead poisoning in trumpeter swans. 11th Trumpeter Swan Society Conference: 153-158.

Degernes, L. A., and R. K. Frank. 1991. Minnesota trumpeter swan mortality, January 1988–June 1989. 12th Trumpeter Swan Society Conference: 111- 113.

Degernes, L. A., P. T. Redig, and M. Freeman. 1991. New treatments for lead poisoned trumpeter swans. 12th Trumpeter Swan Society Conference: 161-162.

Dennington, M. 1988. Trumpeter swan habitat in southern Yukon. 10th Trumpeter Swan Society Conference: 36-41.

Denson, E. P. 1970. The trumpeter swan, *Olor buccinator:* A conservation success and its lessons. *Biological Conservation,* 2: 253-256.

Doyle, T. J. 1994. Expansion of trumpeter swans in the upper Tanana Valley, Alaska. 14th Trumpeter Swan Society Conference: 7-18.

Drewien, R. C., K. R. Clegg, and M. N. Fisher. 1992. Winter capture of trumpeter swans at Harriman State Park, Idaho, and Red Rock Lakes National Wildlife Refuge, Montana. 13th Trumpeter Swan Society Conference: 38-46.

Drewien, R. C., J. T. Herbert, T. W. Aldrich. 2000. Detecting trumpeter swans harvested in tundra swan hunts (Abstract). 17th Trumpeter Swan Society Conference: 155.

Drewien, R. C., J. T. Herbert, T. W. Aldrich, and S. H. Bouffard. 1999. Detecting trumpeter swans harvested in tundra swan hunts. *Wildlife Society Bulletin* 27:95–102.

Drewien, R. C., R. E. Shea, B. Conant, J. S. Hawkings, and N. Hughes. 2004. Satellite tracking trumpeter swans from the Yukon Territory (Abstract). 19th Trumpeter Swan Society Conference: 43.

Ducey, J. E. 1999. History and status of the trumpeter swan in the Nebraska Sand Hills. *North American Swans* 28(1): 31-39.

Eaton, J. 1986. Trumpeter swans at Harriman State Park. 9th Trumpeter Swan Society Conference: 51-52.

Eichholz, M. W. and D. M. Varner. 2007. Survival of Wisconsin Interior Population of trumpeter swans. 20th Trumpeter Swan Society Conference: 45-52.

Engelhardt, K. A. M., J. A. Kadlec, T. W. Aldrich, and V. L. Roy. 1999. The Utah trumpeter swan reintroduction program: proposal to evaluate reintroduction success. 16th Trumpeter Swan Society Conference: 61-65.

Engelhardt, K. A. M., J. A. Kadlec, V. L. Roy, and J. A. Powell. 2000. Evaluation of translocation criteria: Case study with trumpeter swans (*Cygnus buccinator*). *Biological Conservation* 94:173–181.

Eyraud, E. 1986. Harriman State Park: Background, management and trumpeter swans. 9th Trumpeter Swan Society Conference: 48-51.

Fowler, G. M. 1999. Trumpeter swans in the community – Comox Valley, British Columbia. 16th Trumpeter Swan Society Conference: 98-99.

Fowler, G. M., and B. Wareham. 1996. Comox Valley Waterfowl Management Project 1991-94 Report: a report on trumpeter swan management in the Comox Valley, British Columbia. 15th Trumpeter Swan Society Conference: 44-47.

Froelich, A. J., J. C. Johnson, and D. M. Lodge. 1999. Food preferences of mute and trumpeter swans. 16th Trumpeter Swan Society Conference: 133.

Gale, R. S. 1988. Trumpeter swan winter habitat relationships in the Tristate area. 10th Trumpeter Swan Society Conference: 54-56.

Gale, R. S. 1990a. Status of trumpeter swans in Idaho. 11th Trumpeter Swan Society Conference: 4-5.

Gale, R. S. 1990b. Results of the cooperative Rocky Mountain Population trumpeter swan study. 11th Trumpeter Swan Society Conference: 34-37.

Gale, R. S., E. O. Garton, and I. J. Ball. 1987. *The History, Ecology, and Management of the Rocky Mountain Population of Trumpeter Swans*. Missoula, MT: Montana Cooperative Wildlife Research Unit, U.S. Fish and Wildlife Service.

Gillette, L. N. 1986. Status report for the Hennepin County Park Reserve District trumpeter swan restoration project. 9th Trumpeter Swan Society Conference: 94-96.

Gillette, L. N. 1988. Status report for the Hennepin Parks' trumpeter swan restoration project. 10th Trumpeter Swan Society Conference: 104-108.

Gillette, L. N. 1990a. The impact of nesting trumpeter swans on other species of waterfowl. *Trumpeter Swan Society Conference* 11:162–163.

Gillette, L. N. 1990b. Causes of mortality for trumpeter swans in central Minnesota, 1980-1987. 11th Trumpeter Swan Society Conference: 148-151.

Gillette, L. N. 1990c. The impact of nesting trumpeter swans on other species of waterfowl. 11th Trumpeter Swan Society Conference: 162-163.

Gillette, L. N. 1991a. The Trumpeter Swan Society draft position paper on tundra swan hunting. 12th Trumpeter Swan Society Conference: 59-62.

Gillette, L. N. 1991b. Ways to reduce the potential for lead poisoning in trumpeter swans. 12th Trumpeter Swan Society Conference: 119-121.

Gillette, L. N. 1991c. Need for a coordinated restoration approach for the Interiorth Population of trumpeter swans. 12th. Trumpeter Swan Society Conference: 133.

Gillette, L. N. 1991d. Options for establishing migratory populations of Interior Population trumpeter swans. 12th Trumpeter Swan Society Conference: 136-138.

Gillette, L. N. 1996. Building a migratory tradition for the Interior Population of trumpeter swans. 15th Trumpeter Swan Society Conference: 99-103.

Gillette, L. N. 1999. Why is it so hard to establish a migratory population of trumpeter swans? 16th Trumpeter Swan Society Conference: 21-24.

Gillette, L. N. 2000a. What needs to be done to complete the restoration of the Interior Population of trumpeter swans? 17th Trumpeter Swan Society Conference: 35-38.

Gillette, L. N. 2000b. Perspectives of The Trumpeter Swan Society on management of the Rocky Mountain Population of trumpeter swans. 17th Trumpeter Swan Society Conference: 82-84.

Gillette, L. N. 2007. Is migration necessary for restoration of trumpeter swans in the Midwest? 20th Trumpeter Swan Society Conference: 55-57.

Gillette, L. N., and M. H. Linck. 2004. Population status and management options for the Interior Populations of trumpeter swans. 19th Trumpeter Swan Society Conference: 139-147.

Gillette, L. N., and R. Shea. 1995. An evaluation of trumpeter swan management today and a vision for the future. *Transactions of the North American Wildlife and Natural Resources Conference* 60:258–265.

Gomez, D. 2001. 1999. Fall Survey of the Rocky Mountain Population (RMP) of trumpeter swans, U.S. Flocks. 17th Trumpeter Swan Society Conference: 50-52.

Gomez, D., and E. Scheuering. 1996. Termination of artificial feeding at Red Rock Lakes National Wildlife Refuge, Montana. 15th Trumpeter Swan Society Conference: 62-69.

Grant, T., and P. Henson. 1991. Habitat use by trumpeter swans breeding on the Copper River Delta, Alaska. (Abstract). 12th Trumpeter Swan Society Conference: 171.

Groves, D. J. 2012. *The 2010 North American Trumpeter Swan Survey: A Cooperative North American Survey.* Juneau, AK: U.S. Fish and Wildlife Service. 17 pp.

Groves, D. J., Conant, B., and J. I. Hodges. 1997. A summary of Alaska trumpeter swan surveys 1996. *North American Swans* 26(2): 45-49.

Groves, D. J., B. Conant, R. J. King, and D. Logan. 1998. Trumpeter swan surveys on the Chugach National Forest 1997. *North American Swans* 27(1): 36-45.

Groves, D. J., B. Conant, W. W. Larned, and D. Logan. 1999. Trumpeter swan surveys on the Chugach National Forest 1998 – an update. *North American Swans* 28(1): 16-21.

Groves, D. J., B. Conant, E. Mallek, and D. Logan. 2002. Trumpeter swan surveys on the Chugach National Forest 2001 – an update. *North American Swans* 31(1): 19-20.

Groves, D. J., B. Conant, J. Sarvis, and D. Logan. 2001. Trumpeter swan surveys on the Chugach National Forest 2000 – an update. *North American Swans* 30(1): 51-54.

Hammer, D. (moderator). 1986. Panel Discussion: Coordinating management of the Rocky Mountain trumpeter swan population and the role of The Trumpeter Swan Society. 9th Trumpeter Swan Society Conference: 55-61.

Hanauska-Brown, L. A. 2004. Southeastern Idaho trumpeter swan translocations and observations 2001-2003 – project update. 19th Trumpeter Swan Society Conference: 101-107.

Hansen, H. A., Shepherd, P. E. K., King, J. G., & Troyer, W. A. 1971. *The Trumpeter Swan in Alaska.* Wildlife Monographs, no. 26, pp. 1-83.

Hansen, J. L. 1991. Iowa's role in trumpeter swan restoration. 12th Trumpeter Swan Society Conference: 141.

Hawkings, J. S. 1990. Spring staging areas for trumpeter swans in the Southern Lakes Region of Yukon (Abstract). 11th Trumpeter Swan Society Conference: 33.

Hawkings, J. S. 2000. Design and effectiveness of the 1995 Yukon/Northern British Columbia trumpeter swan survey: An appropriate technique for 2000 and beyond? 17th Trumpeter Swan Society Conference: 145-153.

Hawkings, J. S. 2007. The Yukon and northwestern British Columbia trumpeter swan survey, 2005. 20th Trumpeter Swan Society Conference: 61-78.

Hawkings, J. S., and N. L. Hughes. 1994. Recruitment and overwinter survival of Pacific Coast trumpeter swans as determined from age ratio counts. 14th Trumpeter Swan Society Conference: 37-47.

Hemker, T. P. 2004. The trumpeter swan implementation plan – an overview. 19th Trumpeter Swan Society Conference: 134-135.

Henry, A. 2004. Wetland condition on the Caribou–Targhee National Forest, Idaho and Wyoming (Abstract). 19th Trumpeter Swan Society Conference: 120.

Henson, P., and J. A. Cooper. 1992. Division of labour in breeding trumpeter swans *Cygnus buccinator*. *Wildfowl* 43:40–48.

Henson, P., and J. A. Cooper. 1993. Trumpeter swan incubation areas of differing food quality. *Journal of Wildlife Management* 57:709–716.

Henson, P., and T. A. Grant. 1991. The effects of human disturbance on trumpeter swan breeding behavior. *Wildlife Society Bulletin* 19:248–257.

Herbert, J. 1992a. Rocky Mountain Population of trumpeter swans – a Pacific Flyway Study Committee perspective. 13th Trumpeter Swan Society Conference: 19-21.

Herbert, J. 1994. Pacific Flyway experimental "general" swan hunting season – a proposal. 14th Trumpeter Swan Society Conference: 133-136.

Hilliar, C. 1994. The Comox Valley Project Watershed Society (Abstract). 14th Trumpeter Swan Society Conference: 95.

Hills, L. 2004. Spring and fall migration and pond usage by trumpeter swans, Cochrane area, Alberta, Canada, 2002. 19th Trumpeter Swan Society Conference: 60- 70.

Hines, M. E. 1991a. Minnesota DNR trumpeter swan restoration efforts – 1989 status report. 12th Trumpeter Swan Society Conference: 97-99.

Hines, M. E. 1991b. Minnesota DNR efforts – the selection of wetlands for release of 2- year-old trumpeter swans. 12th Trumpeter Swan Society Conference: 100- 104.

Hodges, J. I., B. Conant, and S. L. Cain. 1988. Alaska trumpeter swan 1986 sample, and recommendations for a continent-wide sampling scheme (Abstract). 10th Trumpeter Swan Society Conference: 130.

Hodges, J. I., B. Conant, and S. L. Cain. 1990. A summary of the 1987 Alaska trumpeter swan surveys. 11th Trumpeter Swan Society Conference: 68-71.

Holbek, N. 1994. Tools for dealing with land use problems on the coastal wintering areas, the agricultural land reserve. 14th Trumpeter Swan Society Conference: 89- 92.

Holton, G. 1998. An overview of trumpeter swans in the Grande Prairie region, 1957- 1986. 10th Trumpeter Swan Society Conference: 11-17.

Howie, R. R. 1994. Trumpeter swans wintering in the Thompson-Okanagan areas of British Columbia. 14th Trumpeter Swan Society Conference: 49-60.

Howie, R. R., and R. G. Bison. 2004. Wintering trumpeter and tundra swans in the southern interior of British Columbia. 19th Trumpeter Swan Society Conference: 16-28.

Hughlett, C. A., F. C. Bellrose, H. H. Burgess, A. S. Hawkins, and J. A. Kadlec. 1986. Declining productivity of trumpeter swans at Red Rock Lakes National Wildlife Refuge, Lima, Montana. 9th Trumpeter Swan Society Conference: 124-131.

Innes, D. 1994. Trumpeter swan Pacific Coast Population status in the Comox area of the Vancouver Island, British Columbia. 14th. Trumpeter Swan Society Conference: 72-73.

Ivey, G. L. 1990. Population status of trumpeter swans in southeastern Oregon. *Trumpeter Swan Society Conference* 11:118–122.

Ivey, G. L., and C. G. Carey. 1991. A plan to enhance Oregon's trumpeter swan population. 12th Trumpeter Swan Society Conference: 18-23.

Ivey, G. L., and J. E. Cornely. 2007. Survival analysis of Malheur National Wildlife Refuge trumpeter swans (Abstract). 20th Trumpeter Swan Society Conference: 106.

Ivey, G. L., M. J. St. Louis, and B. D. Bales. 2001. The status of the Oregon trumpeter swan program. 17th Trumpeter Swan Society Conference: 108-113.

Jobes, C. 1990. Growth characteristics of trumpeter swan cygnets from different populations. *Trumpeter Swan Society Conference* 11:164.

Jobes, C. R. 1986. Energetics of growth of trumpeter and mute swan cygnets (Abstract). 9th Trumpeter Swan Society Conference: 119.

Jobes, C. R. 1990. Growth characteristics of trumpeter swan cygnets from different populations (Abstract). 11th Trumpeter Swan Society Conference: 164.

Johnsgard, P. A. 1978 . The triumphant trumpeters. *Natural History*, November, p. 72-77. http://digitalcommons.unl.edu/biosciornithology/18

Johnsgard, P. A. 1982. *Teton Wildlife: Observations by a Naturalist*. Boulder, CO: Colorado Assoc. Univ. Press, 128 pp.

Johnsgard, P. A. 2013. *Yellowstone Wildlife: Ecology and Natural History of the Greater Yellowstone Ecosystem*. Boulder, CO: Univ. Press of Colorado. 239 pp.

Johnson, W. C. (Joe). 1991. Michigan's trumpeter swan restoration program. 12th Trumpeter Swan Society Conference: 108-110.

Johnson, W. C. (Joe). 1999a. Michigan 1996 trumpeter swan update. 16th Trumpeter Swan Society Conference: 18-20.

Johnson, W. C. (Joe). 1999b. Observations of territorial conflict between trumpeter swans and mute swans in Michigan. 16th Trumpeter Swan Society Conference: 134-136.

Johnson, W. C. (Joe). 2007. Status of the Michigan population of trumpeter swans, 2005. 20th Trumpeter Swan Society Conference: 20-21.

Jordan, M. 1986. A summary of the distribution and status of trumpeter swans in Washington State. 9th Trumpeter Swan Society Conference: 67-70.

Jordan, M. 1990. A summary of the status of trumpeter swans in Washington State. 11th Trumpeter Swan Society Conference: 113-116.

Jordan, M. 1991. Trumpeter and tundra swan survey in western Washington and Oregon – January 1989. 12th Trumpeter Swan Society Conference: 14-17.

Jordan, M., L. N. Gillette, R. E. Shea. 2000. Summary of trumpeter swan priorities identified during the 17th Trumpeter Swan Society Conference, September 1999. 17th Trumpeter Swan Society Conference: 179-180.

Kaye, R. and L. Shandruk. 1992. Elk Island National Park trumpeter swan reintroduction – 1990. 13th Trumpeter Swan Society Conference: 22-30.

Killaby, M. 1988. Trumpeter swan habitation and proposed management in Saskatchewan. 10th Trumpeter Swan Society Conference: 47-48.

Kilpatrick, D., K. P. Reese, L. Hanauska- Brown, and T. Hemker. 2007. Trumpeter swan translocation project 2001-2005 in Idaho: Survival and movement (Abstract). 20th Trumpeter Swan Society Conference: 98-99.

Kilpatrick, D. 2007. Translocating trumpeter swans from the Rocky Mountain Population: Habitat, movement, and survival. M.S. thesis, University of Idaho, Moscow, ID.

King, J. G. 1986. Managing to have wild trumpeter swans on a continent exploding with people. 9th Trumpeter Swan Society Conference: 119-123.

King, J. G. 1988. New goals for the second half century of trumpeter swan restoration. 10th Trumpeter Swan Society Conference: 118-120. Lawrence, S. 1999. The Monticello swans. 16th Trumpeter Swan Society Conference: 32-38.

King, J. G. 1994. Pacific Coast trumpeter swans in the 21st century. 14th Trumpeter Swan Society Conference: 3-4.

King, J. G. 1996. Trying to understand what swans think about, especially winter habitat. 15th Trumpeter Swan Society Conference: 41-43.

King, J. G. 2000. Are Alaska's wild swans safe? 17th Trumpeter Swan Society Conference: 4-5.

King, J. G. 2007. Comparison of 290 photos of wild swan nests. 20th Trumpeter Swan Society Conference: 130-135.

King, J. G., R. Ritchie, B. Cooper, and H. McMahan. 1992. Flying with the swans through Alaska's great mountains. 13th Trumpeter Swan Society Conference: 165-168.

King, R. J. 1985. *Trumpeter Swan (Cygnus buccinator) Movements from the Tanana Valley, Alaska*. Fairbanks, AK: U.S. Fish and Wildlife Service.

King, R. J. 1986. Trumpeter swan movements from Tanana Valley, Alaska (Abstract). 9th Trumpeter Swan Society Conference: 65.

King, R. J. 1988. Progress report on impact of collecting trumpeter swan (*Cygnus buccinator*) eggs in Minot Flats, Alaska – 1986. 10th Trumpeter Swan Society Conference: 89-95.

King, R. J. 1990. Impacts on trumpeter swans (*Cygnus buccinator*) from egg collection activities in the Minto Flats, Alaska, 1987 (Abstract). 11th Trumpeter Swan Society Conference: 106.

King, R. J. 1991a. Migration and wintering resightings of trumpeter swans from central Alaska (Abstract). 12th Trumpeter Swan Society Conference: 10.

King, R. J. 1991b. Impacts on trumpeter swans from egg collection activities in Minto Flats, Alaska. 12th Trumpeter Swan Society Conference: 28-44.

King, R. J. 1992. Management problems on the breeding grounds and strategies to resolve them. 13th Trumpeter Swan Society Conference: 160-164.

King, R. J. 1994. Trumpeter swan movements from Minto Flats, Alaska: 1982-92. 14th Trumpeter Swan Society Conference: 19-36.

Kittelson, S. M. 1990a. An update of the Minnesota Department of Natural Resources trumpeter swan restoration project. 11th Trumpeter Swan Society Conference: 50-52.

Kittelson, S. M. 1990b. Avicultural techniques used in Minnesota trumpeter swan restoration. 11th Trumpeter Swan Society Conference: 174-176.

Kittelson, S. M. 1991a. Experiments with wing tags – the third generation design. 12th Trumpeter Swan Society Conference: 53-54.

Kittelson, S. M. 1991b. An update of Minnesota's trumpeter swan restoration efforts – the captive rearing program. 12 Trumpeter Swan Society Conference: 95-96.

Kittelson, S. M. and P. Hines. 1992. Minnesota Department of Natural Resources trumpeter swan restoration project status report. 13th Trumpeter Swan Society Conference: 109-113.

Kraege, D. 1996. Pacific Coast Population – status, trends, and management issues. 15th Trumpeter Swan Society Conference: 3-8.

Kraft, R. H. 1990. Status report of the Lacreek National Wildlife Refuge, South Dakota, trumpeter swan flock management plan. 11th Trumpeter Swan Society Conference: 58-61.

Kraft, R. H. 1991a. Status report of the Lacreek trumpeter swan flock. 12th Trumpeter Swan Society Conference: 88-90.

Kraft, R. H. 1991b. A proposal to restore winter migration in trumpeter swans by establishing breeding pairs in the wintering area. 12th Trumpeter Swan Society Conference: 142.

Kraft, R. H. 1996. Observations of trumpeter swan behavior and management techniques. 15th Trumpeter Swan Society Conference: 91-98.

Kraft, R. H. 2001. Status report of the High Plains Flock for 2000. *North American Swans* 30(1): 32-37.

Kraft, R. H. 2004. Status report of the Lacreek trumpeter swan flock for 2002. 19th Trumpeter Swan Society Conference: 148-154.

Lamb, G. 2004. Swan habitat conservation and partnerships on Long Beach Peninsula, Washington State (Abstract). 19th Trumpeter Swan Society Conference: 9.

LaMontagne, J. M., L. J. Jackson, and R. M. R. Barclay. 2003a. Characteristics of ponds used by trumpeter swans in a spring stopover area. *Canadian Journal of Zoology* 81:1791–1798.

LaMontagne, J. M., L. J. Jackson, and R. M. R. Barclay. 2003b. Compensatory growth responses of *Potamogeton pectinatus* to foraging by migrating trumpeter swans in spring stopover areas. *Aquatic Botany* 76:235–244.

LaMontagne, J. M., L. J. Jackson, and R. M. R. Barclay. 2004. Energy balance of trumpeter swans at stopover areas during spring migration. *Northwestern Naturalist* 85:104–110.

LaMontagne, J. M., R. M. R. Barclay, and L. J. Jackson. 2001. Trumpeter swan behaviour at spring-migration stopover areas in southern Alberta. *Canadian Journal of Zoology* 79: 2036–2042.

Lance, E. W. and E. J. Mallek. 2004. One year of satellite telemetry data for four Alaskan trumpeter swans. 19th Trumpeter Swan Society Conference: 29-38.

Lawrence, S. 1999. The Monticello swans. 16th Trumpeter Swan Society Conference: 32-38.

Lawrence, S. 2007. The trumpeter swans of Monticello, Minnesota. 20th Trumpeter Swan Society Conference: 153-155.

Linck, M. H. 1999. Advantages and disadvantages of a wintering congregation of trumpeter swans on the Mississippi River, Monticello (Wright County), Minnesota. 16th Trumpeter Swan Society Conference: 30-31.

Linck, M. H., K. Rowe, and J. Mosby. 2007. The trumpeter swans of Heber Springs, Cleburne County, Arkansas. 20th Trumpeter Swan Society Conference: 42- 44.

Lockman, D. C. 1986. North American Management Plan as it pertains to the Rocky Mountain Population. 9th Trumpeter Swan Society Conference: 4.

Lockman, D. C. 1988. Wyoming trumpeter swan progress report. 10th Trumpeter Swan Society Conference: 60-63.

Lockman, D. C. 1990a. Wyoming trumpeter swan program status. 11th Trumpeter Swan Society Conference: 9-11.

Lockman, D. C. 1990b. Trumpeter swan mortality in Wyoming, 1982-1987. 11 Trumpeter Swan Society Conference: 12-13.

Lockman, D. C. 1990c. Rocky Mountain Trumpeter Swan Population Subcommittee report (Abstract). 11th Trumpeter Swan Society Conference: 38.

Lockman, D. C. 1990d. Rocky Mountain Trumpeter Swan Population range expansion project, 1988-1993. 11th Trumpeter Swan Society Conference: 40-43.

Lockman, D. C. 1991. Strategies tested in Wyoming for trumpeter swan range expansion – 1989 progress report. 12 Trumpeter Swan Society Conference: 196-199.

Lockman, D. C., C. D. Mitchell, B. Reiswig, and R. S. Gale. 1990. Identifying potential winter habitat for trumpeter swans. 11th Trumpeter Swan Society Conference: 20-22.

Lockman, D. C., R. Wood, H. H. Burgess, R. Burgess, and H. Smith. 1990. Trumpeter swan seasonal habitat use in western Wyoming. 11th Trumpeter Swan Society Conference: 14-19.

Lockman, D. C., R. Wood, B. Smith, B. Raynes, and D. Childress. 1986. Progress report: Rocky Mountain Trumpeter Swan Population – Wyoming flock, 16 September 1983 through 15 September 1984. 9th Trumpeter Swan Society Conference: 29-47.

Long, W. M., and D. Stevenson. 1999. Trumpeter swan range restoration in Wyoming. 16th Trumpeter Swan Society Conference: 66-71.

Loranger, A., and D. Lons. 1990. Relative abundance of sympatric trumpeter and tundra swan populations in west-central interior Alaska. 11th Trumpeter Swan Society Conference: 92-98.

Luebbert, J. 2000. The science of migration and navigation: Considerations for trumpeter swan (*Cygnus buccinator*) translocations. 17th Trumpeter Swan Society Conference: 164-169.

Lueck, C. M. 1991. The agonistic behavior of nesting trumpeter swans toward other waterfowl. 12th Trumpeter Swan Society Conference: 163-169.

Lumbis, K. 1988. Wetland habitat management in the Grande Prairie region of northwestern Alberta. 10th Trumpeter Swan Society Conference: 22-27.

Lumsden, H. G. 1984. The pre-settlement breeding distribution of trumpeter, *Cygnus buccinator*, and tundra swans, *C. columbianus columbianus,* in eastern Canada. *Canadian Field-Naturalist* 98:415–424.

Lumsden, H. G. 1986. The trumpeter swan/mute swan experiment: Ontario. 9th Trumpeter Swan Society Conference: 117-118.

Lumsden, H. G. 1988a. Productivity of trumpeter swans in relation to condition. 10th Trumpeter Swan Society Conference: 150-154.

Lumsden, H. G. 1988b. The food of trumpeter swan cygnets in Ontario. 10th Trumpeter Swan Society Conference: 155- 157.

Lumsden, H. G. 1990. The trumpeter swan restoration program in Ontario, 1987 progress report. 11th Trumpeter Swan Society Conference: 48-49.

Lumsden, H. G. 1991. Ontario trumpeter swan restoration program – progress report 1989. 12th Trumpeter Swan Society Conference: 105.

Lumsden, H. G. 1992. Ontario trumpeter swan restoration program progress report- 1990. 13th Trumpeter Swan Society Conference: 119-122.

Lumsden, H. G. 1997. The trumpeter swan restoration program in Ontario - 1 September 1997. *North American Swans* 26(2): 42-44.

Lumsden, H. G. 1999a. The trumpeter swan restoration program in Ontario 1998. *North American Swans* 28(1): 22-24.

Lumsden, H. G. 1999b. Pair formation in captive trumpeter swans. 16th Trumpeter Swan Society Conference: 117- 121.

Lumsden, H.G. 1999c. Trumpeter swan restoration in Ontario. 16th Trumpeter Swan Society Conference: 10-13.

Lumsden, H. G. 2000a. The trumpeter swan restoration program in Ontario – 1999. 17th Trumpeter Swan Society Conference: 11-15.

Lumsden, H. G. 2000b. Induced migration - its origins and history. 17th Trumpeter Swan Society Conference: 158-162.

Lumsden, H. G. 2001. A survey of trumpeter swans in the Kenora District of Ontario. North American Swans 30(1): 19-20.

Lumsden, H. G. 2002a. Laying and incubation behavior of captive trumpeter swans. *Waterbirds* 25 (Special Publication 1): 293–295.

Lumsden, H. G. 2002b. Hatchability of eggs from captive trumpeter swans. *North American Swans* 31(1): 2-4.

Lumsden, H. G. 2002c. The trumpeter swan restoration program in Ontario — 2001. *North American Swans* 31(1): 16-18.

Lumsden, H. G. 2004. The trumpeter swan restoration program in Ontario — 2002. 19th Trumpeter Swan Society Conference: 155-158.

Lumsden, H. G. 2007a. The inventory of trumpeter swans in Ontario in 2005. 20th Trumpeter Swan Society Conference: 11.

Lumsden, H. G. 2007b. Migration of Ontario trumpeter swans. 20th Trumpeter Swan Society Conference: 37-40.

Lumsden, H. G., D. Compton, J. Johnson, S. Kittelson, P. Hines, S. Matteson, and J. Smith. 1994. Trumpeter swan restoration in the Midwest. 14th Trumpeter Swan Society Conference: 145-149.

Lumsden, H. G., and M. C. Drever. 2002. Overview of the trumpeter swan reintroduction program in Ontario, 1982–2000. *Waterbirds* 25 (Special Publication 1): 301–312.

Lumsden, H. G., D. McLachlin, and P. Nash. 1988. Restoration of trumpeter swans in Ontario. 10th Trumpeter Swan Society Conference: 110-116.

Luszcz, D. 2000. Status of Atlantic Flyway Trumpeter Swan Management Plan. 17th Trumpeter Swan Society Conference: 9-10.

Mackay, J. 2004. The Ruby Lake trumpeter swan flock: Its history, current status, and future. 19th Trumpeter Swan Society Conference: 121-127.

Mackay, R. H. 1988. Trumpeter swan investigations, Grand Prairie, Alberta, 1953–75. *Trumpeter Swan Society Conference* 10:5–10..

Maj, M. 1986. Targhee National Forest trumpeter swans. 9th Trumpeter Swan Society Conference: 52-53.

Marsolais, J. V. 1994. The genetic status of trumpeter swan (*Cygnus buccinator*) populations. (Abstract). 14th Trumpeter Swan Society Conference: 162-164.

Mascall, M. 1994. West Coast Islands' Stewardship and Conservancy Society of British Columbia. 14th Trumpeter Swan Society Conference: 96-97.

Matteson, S. W. 1991. Wisconsin's trumpeter swan recovery program. 12th Trumpeter Swan Society Conference: 106- 107.

Matteson, S., S. Craven and D. Compton. 1995. *The Trumpeter Swan.* Madison, WI: University of Wisconsin Extension, Publication G3647.

Matteson, S. W., P. F. Manthey, M. J. Mossman, and L. M. Hartman. 2007. Wisconsin trumpeter swan recovery program: Progress toward restoration, 1987–2005. *Trumpeter Swan Society Conference* 20:11–18.

Matteson, S. W., M. J. Mossman, and L. M. Hartman. 1996. Wisconsin's trumpeter swan restoration efforts, 1987-1994. 15th. Trumpeter Swan Society Conference: 73-85.

REFERENCES

Matteson, S. W., M. J. Mossman, R. Jurewicz, and E. Diebold. 1992. Collection, transport, and hatching success of Alaskan trumpeter swan eggs 1989-90, and status of Wisconsin's trumpeter swan recovery program. 13th Trumpeter Swan Society Conference: 105-108.

McCormick, K. J. 1986. Status of trumpeter swans in the Northwest Territories. 9th Trumpeter Swan Society Conference: 22-27.

McEneaney, T. 1986. Effects of water flow fluctuations, icing, and recreationists on the distribution of wintering trumpeter swans in the Tristate region. 9th Trumpeter Swan Society Conference: 53-54.

McEneaney, T. 1991. Status of the trumpeter swan in Yellowstone National Park. 12th Trumpeter Swan Society Conference: 193-194.

McEneaney, T. 1996. Trumpeter swan management within and beyond park boundaries. 15th Trumpeter Swan Society Conference: 151-155.

McEneaney, T. 2005. Rare color variants of the trumpeter swan. *Birding* 37:148–154.

McKelvey, R. W. 1986a. An overview of the North American Management Plan for trumpeter swans as it pertains to the Pacific Coast Population. 9th Trumpeter Swan Society Conference: 62-64.

McKelvey, R. W. 1986b. Notes on the status of trumpeter swans in British Columbia and the Yukon Territory, and on grazing studies at Comox Harbour, British Columbia. 9th Trumpeter Swan Society Conference: 65-67.

McKelvey, R. W. 1986c. Guidelines for trumpeter swan restoration in Canada. 9th Trumpeter Swan Society Conference: 111-112.

McKelvey, R. W. 1988. The 1985 survey of trumpeter swans in British Columbia and Yukon. 10th Trumpeter Swan Society Conference: 145.

McKelvey, R. W. 1991. The status of trumpeter swans wintering in southwestern British Columbia in 1989. 12th Trumpeter Swan Society Conference: 12-13.

McKelvey, R. W. 1992. Canadian involvement in management of trumpeter swans. 13th Trumpeter Swan Society Conference: 173-175.

McKelvey, R. W., and N. A. M. Verbeek. 1988. Habitat use, behaviour and management of trumpeter swans, *Cygnus buccinator*, wintering at Comox, British Columbia. *Canadian Field-Naturalist* 102:434–441.

Mitchell, C. D. 1990. Efficiency of techniques for feeding wintering trumpeter swans. 11th Trumpeter Swan Society Conference: 170-173.

Mitchell, C. D. 1991. Update on trumpeter swans in Montana – 1988-89. 12th Trumpeter Swan Society Conference: 189-192.

Mitchell, C. D. 1994. Trumpeter swan research needs. 14th Trumpeter Swan Society Conference: 155-161.

Mitchell, C. D. 2004. Trumpeter swan restoration at Grays Lake National Wildlife Refuge, Idaho. 19th Trumpeter Swan Society Conference: 108-115.

Mitchell, C. D., and M. W. Eichholz. 1994. Trumpeter swan (*Cygnus buccinator*). T*he Birds of North America* 105. The Birds of North America, Inc. Philadelphia, PA: The Academy of Natural Sciences, and Washington, DC: American Ornithologists' Union, 24 pp.

Mitchell, C. D., and J. J. Rotella. 1997. Brood amalgamation in trumpeter swans. *Wildfowl* 48:1–5.

Mitchell, C. D., and L. Shandruk. 1992. Rocky Mountain Population of trumpeter swans: Status, distribution, and movements. 13th Trumpeter Swan Society Conference: 3-18.

Monnie, J. B. 1966. Reintroduction of the trumpeter swan to its former breeding range. *Journal of Wildlife Management* 30: 691–696.

Monson, M. A. 1956. Nesting of trumpeter swans in the lower Copper River basin, Alaska. *Condor* 58:444–445.

Moriarty, J. J. 1991. Feather stress in trumpeter swans. 12th Trumpeter Swan Society Conference: 152.

Morrison, K. F. 1990. Number and age composition of trumpeter swans wintering on the east coast of Vancouver Island, British Columbia, 1983-1988. 11th Trumpeter Swan Society Conference: 107-112.

Mossman, M. J., and S. W. Matteson. 1990. Trumpeter swan status report for Wisconsin. 11th Trumpeter Swan Society Conference: 53-55.

Munoz, R. 2000. Role of Southeast Idaho National Wildlife Refuge Complex in the Rocky Mountain Population trumpeter swan project. 17th Trumpeter Swan Society Conference: 107-108.

Murphy, S., R. Rey, and P. R. Demaris. 1996. Rehabilitation and research on Trumpeter and tundra swans with lead poisoning in Washington's Skagit Valley area. 15th Trumpeter Swan Society Conference: 117-130.

Myhr, R. 1996. Preserving trumpeter swan habitat in the San Juan Islands: Perhaps an example for other land trusts. 15th Trumpeter Swan Society Conference: 136-138.

Nelson, H. K. 1994. The Trumpeter Swan Society's position on swan hunting. 14th Trumpeter Swan Society Conference: 137-142.

Nelson, H. K. 1999. Development of a management plan for the Interior Population of trumpeter swans. 16th Trumpeter Swan Society Conference: 27-29.

Neptune, B. 2007. The nesting trumpeter swans of Dawn, Missouri. 20th Trumpeter Swan Society Conference: 158-159.

Niethammer, K. R., D. Gomez, and S. M. Linneman. 1994. Termination of winter feeding of trumpeter swans at Red Rock Lakes National Wildlife Refuge – a progress report. 14th Trumpeter Swan Society Conference: 118-121.

North, M. R. 2007. The earliest historical records of trumpeter swans – extralimital to today's distribution. 20th Trumpeter Swan Society Conference: 148-150.

Norton, M. and G. Beyersbergen. 2001 2000 survey of trumpeter swans in Alberta, Saskatchewan, Manitoba, and the Northwest Territories. *North American Swans* 30(1): 25-31.

Olson, D. 2001. 2001 Midwinter survey: Rocky Mountain Population of trumpeter swans. *North American Swans* 30(1): 38-42.

Olson, D. 2002. 2002 Midwinter survey: Rocky Mountain Population of trumpeter swans. *North American Swans* 31(1): 12-15.

Olson, D. 2004. Analysis of winter satellite telemetry locations from trumpeter swans marked and released at Red Rock lakes National Wildlife Refuge, Montana (Abstract). 19th Trumpeter Swan Society Conference: 95.

Olson, D. 2016. Trumpeter Swan Survey of the Rocky Mountain Population, U.S. Breeding Segment. Fall 2015. Lakewood, CO: U.S. Fish and Wildlife Service, Mountain Prairie Region. 39 pp.

Olson, D., and R. Gregory. 2007. Status of trumpeter swans (*Cygnus buccinator*) at Seney National Wildlife Refuge (1991- 2005) (Abstract). 20th Trumpeter Swan Society Conference: 22.

Orme, M. L., and R. E. Shea. 2000. Trumpeter swan nesting habitat on the Targhee National Forest. 17th Trumpeter Swan Society Conference: 95-102.

Oyler-McCance, S. J., and T. W. Quinn. 2004. Comparison of trumpeter swan populations using nuclear and mitochondrial genetic markers (Abstract). 19th Trumpeter Swan Society Conference: 119.

Pacific Flyway Council. 2002. *Pacific Flyway Implementation Plan for the Rocky Mountain Population of Trumpeter Swans.* Portland, OR: Pacific Flyway Study Committee, U.S. Fish and Wildlife Service.

Page, R. D. 1976. *The Ecology of the Trumpeter Swan on Red Rock Lakes National Wildlife Refuge, Montana.* Ph.D. dissertation, University of Montana, Missoula, MT.

Patla, S., and R. Oakleaf. 2004. Summary and update of trumpeter swan range expansion efforts in Wyoming, 1988-2003. 19th Trumpeter Swan Society Conference 116-118.

Patton, K., E. Butterworth, D. Falk, A. Leach, and C. Smith. 2004. Records of trumpeter swans in the Ducks Unlimited Canada western boreal program. 19th Trumpeter Swan Society Conference: 44-49.

Pelizza, C. A. 2000. Winter site selection characteristics, genetic composition and mortality factors of the High Plains flock of trumpeter swans. 17th Trumpeter Swan Society Conference: 29-34.

Pichner, J. 1991. Trumpeter swan multiple and continuous clutching – a summary. 12th Trumpeter Swan Society Conference: 148-151.

Pichner, J., S. Kittelson, and P. Hines. 1992. Survival of hand-reared and parent-reared trumpeter swans (*Cygnus buccinator*) in the Minnesota Department of Natural Resources restoration project. 13th Trumpeter Swan Society Conference: 114-118.

Pichner, J., N. Reindl, and B. Geiszler. 1990. Double clutching of trumpeter swans (*Cygnus cygnus buccinator*) at the Minnesota Zoological Garden. 11th Trumpeter Swan Society Conference: 177-178.

Price, A. L., J. Kyler, and R. L. Studebaker. 1999. Public participation in the restoration of the trumpeter swans within the Interior Population. 16th Trumpeter Swan Society Conference: 39-43

Proffitt, K. M., T. P. McEneaney, P. J. White, and R. A. Garrott. 2009. Trumpeter swan abundance and growth rates in Yellowstone National Park. *Journal of Wildlife Management* 73:728–736.

Proffitt, K. M., T. P. McEneaney, P. J. White, and R. A. Garrott. 2010. Productivity and fledging success of trumpeter swans in Yellowstone National Park, 1987–2007. *Waterbirds* 33: 341–348.

Rasmussen, P. J. 2007. Managing Monticello trumpeter swans and power line issues (Abstract). 20th Trumpeter Swan Society Conference: 34-36.

Reiswig, B. 1986. Status of the Tristate subpopulation and the Rocky Mountain winter population of trumpeter swans. 9th Trumpeter Swan Society Conference: 7-10.

Reiswig, B. 1988a. The Trumpeter Swan Society's Red Rock Lakes Study Committee recommendations: A U.S. Fish and Wildlife Service update. 10th Trumpeter Swan Society Conference: 50- 53.

Reiswig, B. 1988b. A review of wintering Rocky Mountain trumpeter swan population survey estimates: 1977-1986. 10th Trumpeter Swan Society Conference: 57-59.

Reiswig, B. 1990. Trumpeter swan management – Montana overview. 11th. Trumpeter Swan Society Conference: 2-3.

Reiswig, B., and C. D. Mitchell. 1996. Rocky Mountain Population of trumpeter swans: status, trends, problems, outlook. 15th Trumpeter Swan Society Conference: 9-17.

Ripley, L. 1984. *Alberta/B.C. Cooperative Trumpeter Swan Restoration Program, 1983.* Brooks, AB: Alberta Department of Forestry, Lands and Wildlife,

Ripley, L. 1985. *Trumpeter Swan Restoration Program, 1984.* Brooks, AB: Alberta Department of Forestry, Lands and Wildlife.

Rocky Mountain Population Trumpeter Swan Subcommittee. 1990. Contingency plan for management of wintering trumpeter swans in the vicinity of Harriman State Park, Idaho. 11th Trumpeter Swan Society Conference: 44-45.

Rogers, P. M., and D. A. Hammer. 1998. Ancestral breeding and wintering ranges of

Rolston, G. 1994. Land use conflicts in the Comox Valley. 14th Trumpeter Swan Society Conference: 77-78.

Roy, V. 1996. Trumpeter and tundra swans: Their history and future at the Bear River Migratory Bird Refuge. 15th Trumpeter Swan Society Conference: 53-61.

Schmidt, J., M. S. Lindberg, D. S. Johnson, B. Conant and J. G. King. 2007. Factors affecting the growth and distribution of trumpeter swan populations in Alaska from 1968-2005 (Preliminary results) (Abstract). 20th Trumpeter Swan Society Conference: 113.

Schmidt, P. 2000. Challenges in conserving swans and other migratory birds into the next millennium. 17th Trumpeter Swan Society Conference: 41-43.

Shandruk, L. J. 1986. Draft proposal: A long range habitat management strategy for the Interior Canada subpopulation of trumpeter swans. 9th Trumpeter Swan Society Conference: 14-22.

Shandruk, L. J. 1988a. Status of trumpeter swans in the southern Mackenzie District, Northwest Territories. 10th Trumpeter Swan Society Conference: 42-46.

Shandruk, L. J. 1988b. Elk Island National Park trumpeter swan transplant pilot project – final report. 10th Trumpeter Swan Society Conference: 66-77.

Shandruk, L. J. 1988c. A survey of trumpeter swan breeding habitats in Alberta, Saskatchewan, and northeastern British Columbia. 10th Trumpeter Swan Society Conference: 131-144.

Shandruk, L. J. and G. Holton. 1986. Status report: A pilot project to transplant trumpeter swans into Elk Island National Park, Alberta. 9th Trumpeter Swan Society Conference: 11-13.

Shandruk, L. J. and R. Kaye. 1991. Elk Island National Park trumpeter swan reintroduction – 1989 progress report. 12th Trumpeter Swan Society Conference: 184-188.

Shandruk, L. J. and K. J. McCormick. 1990. Status of trumpeter swans in the southern Mackenzie District, Northwest Territories, 1986 and 1987. 11th Trumpeter Swan Society Conference: 23-27.

Shandruk, L. J., and K. J. McCormick. 1991. Status of the Grande Prairie and Nahanni trumpeter swan flocks. 12th Trumpeter Swan Society Conference: 181- 183.

Shandruk, L. J., and T. Winkler. 1990. Elk Island National Park trumpeter swan reintroduction, 1987 progress report. 11th. Trumpeter Swan Society Conference: 28-32.

Sharp, D. E., and J. B. Bortner. 1991. North American Waterfowl Management Plan. 12th Trumpeter Swan Society Conference: 122-126.

Shea, R. E. 1992. Response of trumpeter swans to trapping at Red Rock Lakes National Wildlife Refuge, Montana, and Harriman State Park, Idaho, winter 1990- 91. 13th Trumpeter Swan Society Conference: 31-37

Shea, R. E. 1999. Recent changes in distribution and abundance of the Rocky Mountain Population of trumpeter swans. 16th Trumpeter Swan Society Conference: 49-55.

Shea, R. E. 2000. Rocky Mountain trumpeter swans: current vulnerability and restoration potential. 17th Trumpeter Swan Society Conference: 74-81.

Shea, R. E. 2004. Status of trumpeter swans nesting in the western United States and management issues. 19th Trumpeter Swan Society Conference: 85-94.

Shea, R. E., R. C. Drewien, and C. S. Peck. 1994. Overview of efforts to expand the range of the Rocky Mountain Population of trumpeter swans. 14th Trumpeter Swan Society Conference: 111-117.

Sheehan, B. 1988. Early history of trumpeter swans in the Grande Prairie area. 10th Trumpeter Swan Society Conference: 2-4.

Shepherd, P. E. K. 1962. An ecological reconnaissance of the trumpeter swan in south central Alaska. M.S. thesis, Washington State University, Pullman, WA.

Shields, R. 1986. Management of the Tristate Swan subpopulation. 9th Trumpeter Swan Society Conference: 5-6.

Sladen, W. J. L., and R. J. Limpert. 1992. A new look at the coded color neck and tarsus band protocol for North American swans. 13th Trumpeter Swan Society Conference: 92-101.

Sladen, W. J. L., and G. H. Olsen. 2007. Teaching geese, swans and cranes pre- selected migration routes using ultralight aircraft, 1990-2004 – looking into the future. 20th Trumpeter Swan Society Conference: 53-54.

Sladen, W. J. L., and D. L. Rininger. 2000. Teaching trumpeter swans pre-selected migration routes using ultralight aircraft as surrogate parents – second experiment, 1998-1999. 17th Trumpeter Swan Society Conference: 163-165.

Sladen, W. J. L., and J. C. Whissel. 2007. The winter distribution of trumpeter swans in relation to breeding areas: the first neckband study, 1972-81. 20th Trumpeter Swan Society Conference: 117- 125.

Slater, G. L. 2006. *Trumpeter Swan (Cygnus buccinator): A Technical Conservation Assessment.* Ecostudies Institute, Mount Vernon, WA: Report prepared for the Species Conservation Project, U.S. Forest Service, Rocky Mountain Region.

Smith, C. S. 1996. Pacific Coast Joint Venture projects that secure or enhance swan habitat. 15th Trumpeter Swan Society Conference: 133-135.

Smith, D. W., and N. Chambers. 2011. *The Future of Trumpeter Swans in Yellowstone National Park: Final Report Summarizing Expert Workshop, April 26–27, 2011.* Yellowstone National Park, WY: Yellowstone Center for Resources, National Park Service,

Smith, J. W. 1988. Status of Missouri's experimental trumpeter swan restoration program. 10th Trumpeter Swan Society Conference: 100-103.

Smith, J. W. 1990. Trumpeter swan status report for Missouri. 11th Trumpeter Swan Society Conference: 56-57.

Smith, J. W., and J. D. Wilson. 1986. Experimental restoration of trumpeter swans to Missouri. 9th Trumpeter Swan Society Conference: 112-115.

Smith, M. C., J. M. Grassley, C. E. Grue, M. Davison, J. Bohannon, C. Schneider and L. Wilson. 2007. Mortality of swans due to ingestion of lead shot, Whatcom County, Washington, and Sumas prairie, British Columbia. 20th Trumpeter Swan Society Conference: 114-116.

Snyder, J. W. 1990. Wintering and foraging ecology of the trumpeter swan, Harriman State Park, Idaho. 11th Trumpeter Swan Society Conference: 6-8.

Snyder, J. W. 1991a. Trumpeter swan winter habitat use on the Henry's Fork. 12th Trumpeter Swan Society Conference: 174.

Snyder, J. W. 1991b. The wintering and foraging ecology of the trumpeter swan, Harriman State Park of Idaho. M.S. thesis, Idaho State University, Pocatello, ID.

Sojda, R. S., J. E. Cornely, L. H. Fredrickson, and A. E. Howe. 2000. Current research efforts in decision support system technology as applied to trumpeter swan management. 17th Trumpeter Swan Society Conference: 139-144.

Sojda, R. S., J. E. Cornely, and A. E. Howe. 2002. Development of an expert system for assessing trumpeter swan breeding habitat in the northern Rocky Mountains. *Waterbirds* 25 (Special Publication 1):313–318.

Squires, J.R. 1991a. The movements, productivity, and habitat-use patterns of trumpeter swans in the Greater Yellowstone Area. 12th Trumpeter Swan Society Conference: 172-173.

Squires, J. R. 1991b. Trumpeter swan food habits, forage processing, activities, and habitat use. Ph.D. dissertation, University of Wyoming, Laramie, WY.

Squires, J. R., and S. H. Anderson. 1995. Trumpeter swan (*Cygnus buccinator*) food habits in the greater Yellowstone ecosystem. *American Midland Naturalist* 133:274–282.

Squires, J. R., and S. H. Anderson. 1997. Changes in trumpeter swan (*Cygnus buccinator*) activities from winter to spring in the greater Yellowstone area. *American Midland Naturalist* 138:208–214.

St. Louis, M. J. 1994. Status of Oregon's trumpeter swan program. 14th Trumpeter Swan Society Conference: 122- 130.

Stearns, F. D., S. Breeser, and D. Sowards. 1990. Population expansion of trumpeter swans in the upper Tanana Valley, Alaska, 1982-1987. 11th Trumpeter Swan Society Conference: 99-105.

Strong, F. E. 1994. Living with swans. 14th Trumpeter Swan Society Conference: 83.

Tessman, S. A. 2000. Pacific Flyway Study Committee perspective on RMP [Rocky Mountain Population] trumpeter swan restoration. 17th Trumpeter Swan Society Conference: 67-73.

Tori, G. M. 1999. Ohio's trumpeter swan restoration project – first year summary. 16th Trumpeter Swan Society Conference: 14-17.

Trost, R. E., J. E. Cornely, and J. B. Bortner. 2000. U.S. Fish and Wildlife Service Perspective on RMP [Rocky Mountain Population] trumpeter swan restoration. 17th Trumpeter Swan Society Conference: 60-66.

Turner, B. 1988. Summary of results of Grande Prairie trumpeter swan collaring program. (Abstract). 10th Trumpeter Swan Society Conference: 28.

Turner, B. C., and R. H. Mackay. 1982. *The Population Dynamics of the Trumpeter Swans of Grand Prairie, Alberta.* Edmonton, AB: Unpublished report, Canadian Wildlife Service.

Van Kirk, R., and R. Martin. 2000. Interactions among waterfowl herbivory, aquatic vegetation, fisheries and flows below Island Park Dam, Idaho. 17th Trumpeter Swan Society Conference: 85-94.

Vos, A. de 1964. Observations on the behavior of captive trumpeter swans during the breeding season. *Ardea,* 52: 166-189.

Vrtiska, M. P., and S. Comeau. 2009. *Trumpeter Swan Survey of the High Plains Flock, Interior Population.* Lincoln, NE: Nebraska Game and Parks Commission.

Vrtiska, M. P., J. L. Hansen, and D. E. Sharp. 2007. Central Flyway perspectives on trumpeter swans. *Trumpeter Swan Society Conference* 20:29–33.

Wareham, W., and G. Fowler. 1994. The Comox Valley waterfowl management project, 1991-93. 14th Trumpeter Swan Society Conference: 93-94.

White, M. 2004. Habitat and management trends affecting trumpeter swans in Alberta. 19th Trumpeter Swan Society Conference: 70-81.

White, M., and R. White. 2000. Rocky Mountain Population of trumpeter swans: Habitat trends in the Grande Prairie region. 17th Trumpeter Swan Society Conference: 128-133.

Whitman, C. L., and C. D. Mitchell. 2004. Winter trumpeter swan mortality in southwestern Montana, eastern Idaho, and northwestern Wyoming, November 2000 through January 2003. 19th Trumpeter Swan Society Conference: 96-100.

Wilk, R. J. 1993. Observations on sympatric tundra, *Cygnus columbianus,* and trumpeter swans, *C. buccinator,* in north-central Alaska, 1989–1991. *Canadian Field-Naturalist* 107: 64–68.

Will, G. C. 1991. Status of the Rocky Mountain Population of trumpeter swans, the range expansion project, and conditions in Idaho. 12th Trumpeter Swan Society Conference: 177-180.

WHOOPER SWAN

Boyd, H., and S. K. Eltringham. 1962. The whooper swan in England. *Bird Study* 9:227-241.

Brazil, M. A. 1981. Geographical variation in the bill patterns of whooper swans. *Wildfowl* 32: 129-131.

Einarsson, Á. 1996. Breeding biology of the whooper swan and factors affecting its breeding success, with notes on its social dynamics and life cycle in the wintering range. Ph.D. dissertation, University of Bristol, Bristol, UK.

Haapenen, A., M. Helminen, and H. K. Soumalainen. 1973a. The spring arrival and breeding phenology of the whooper swan *Cygnus c. cygnus* in Finland. *Finnish Game Research* 33:3–38.

Haapenen, A., M. Helminen, and H. K. Soumalainen. 1973b. Population growth and breeding biology of the whooper swan *Cygnus c. cygnus* in Finland in 1950–1970. *Finnish Game Research* 33:39–60.

Haapenen, A., M. Helminen, and H. K. Soumalainen. 1977. The summer behavior and habitat use of the whooper swan. *Finnish Game Research* 36:49–81.

Kenyon, K. 1961. Birds of Amchitka Island, Alaska. *Auk* 78: 305-325.

Kenyon, K. 1963. Further observations of whooper swans in the Aleutian Islands. *Auk* 80: 540-542.

Laubek, B., L. Nilsson, M. Wieloch, K. Koffijberg, C. Sudfelt, and A. Follestad. 1999. Distribution, number, and habitat choice of the Northwestern European whooper swan (*Cygnus cygnus*) population: results of an international whooper swan census January 1995. *Vögelwelt* 120:141–150.

Rees, E. C., J. M. Black, C. J. Spray, and S. Thorisson, 1991. Comparative study of the breeding success of whooper swans, *Cygnus cygnus*, nesting in upland and lowland regions of Iceland. *Ibis* 133: 365–373.

Rees, E C., J. S. Kirby, and A. Gilburn. 1997. Site selection by swans wintering in Britain and Ireland: the importance of geography, location and habitat. *Ibis* 139:337–352.

TUNDRA (WHISTLING) SWAN

Babcock, C. A., A. C. Fowler, and C. R. Ely. 2002. Nesting ecology of tundra swans on the coastal Yukon-Kuskokwim Delta, Alaska. *Waterbirds* 25 (Special Publication 1): 236–240.

Badzinski, S. S. 2003. Dominance relations and agonistic behaviour of tundra swans (*Cygnus columbianus columbianus*) during fall and spring migration. *Canadian Journal of Zoology* 81:727–733..

Badzinski, S. S. 2005. Social influences on tundra swan activities during migration. *Waterbirds* 28:316–325.

Bart, J., R. Limpert, S. L. Earnst, W. Sladen, J. Hines, and T. Rothe. 1991b. Demography of Eastern Population tundra swans *Cygnus columbianus columbianus*. *International Swan Symposium* 3:178–184.

Bart, J. and J. D. Nichols. 1992. Movements of tundra swans on the East Coast in winter. (Abstract). 13th Trumpeter Swan Society Conference: 52.

Bartonek, J. C., J. R. Serie, and K. A. Converse. 1991. Mortality in tundra swans *Cygnus columbianus*. *International Swan Symposium* 3:356–358.

Bortner, J. B. 1988. Bioenergetics of wintering tundra swans in the Mattamuskeet region of North Carolina (Abstract). 10th Trumpeter Swan Society Conference: 158.

Boyd, S. 1994. Abundance patterns of trumpeter swans and tundra swans on the Fraser River Delta, British Columbia (Abstract). 14th Trumpeter Swan Society Conference: 48.

Canniff, R. S. 1990. Trumpeter and tundra swan collar sightings in the Skagit Valley, 1977-1978 to 1987-1988. 11th Trumpeter Swan Society Conference: 125- 141.

Central Flyway Council. 1991. Position statement on tundra swan hunting in the Central Flyway relative to potential conflicts with trumpeter swan restoration. 12th Trumpeter Swan Society Conference: 73-74.

Childress, D. 1991. Pacific Flyway Council comments on the draft position statement on tundra swan hunting. 12th Trumpeter Swan Society Conference: 75.

Cooper, B. A., and R. J. Ritchie. 1990. Migration of trumpeter and tundra swans in east-central Alaska during spring and fall, 1987. 11th Trumpeter Swan Society Conference: 82-91.

Crawley, D. R., and E. G. Bolen. 2002. Effect of tundra swan grazing in winter wheat in North Carolina. *Waterbirds* 25 (Special Publication 1):162–167.

REFERENCES

Drewien, R. C., J. T. Herbert, T. W. Aldrich. 2000. Detecting trumpeter swans harvested in tundra swan hunts (Abstract). 17th Trumpeter Swan Society Conference: 155

Earnst, S. L. 1992a. Behavior and ecology of tundra swans during summer, autumn, and winter. Ph.D. dissertation, Ohio State University, Columbus, OH.

Earnst, S. L. 1992b. The timing of wing molt in tundra swans: Energetic and non-energetic constraints. *Condor* 94:847–856.

Earnst, S. L. 1992c The habitat use of tundra swans (*Cygnus columbianus columbianus*) on an autumn migratory stopover (Abstract). 13th Trumpeter Swan Society Conference: 51.

Earnst, S. L. 1994. Tundra swan habitat preferences during migration in North Dakota. *Journal of Wildlife Management* 58:546–551.

Earnst, S. L. 2002. Parental care in tundra swans during the pre-fledging period. *Waterbirds* 25:268–277.

Earnst, S. L., and J. Bart. 1991. Costs and benefits of extended parental care in tundra swans, *Cygnus columbianus columbianus*. *International Swan Symposium* 3:260–267.

Earnst, S. L., and T. C. Rothe. 1994. Habitat preferences of tundra swans on their breeding grounds in northern Alaska (Abstract). 14th Trumpeter Swan Society Conference: 177.

Evans, M. E., and W. J. L. Sladen. 1980. A comparative analysis of the bill markings of whistling and Bewick's swans and out-of-range occurrences of the two taxa. *Auk* 97:697–703.

Gillette, L. N. 1992a. Position paper on tundra swan hunting: Introductory remarks. 13th Trumpeter Swan Society Conference: 53-55.

Gillette, L. N. 1992b. Potential techniques for monitoring the harvest in tundra swan hunts. 13th Trumpeter Swan Society Conference: 81-84.

Hawkins, L. L. 1986a. Tundra swan (*Cygnus columbianus columbianus*) breeding behavior. M.S. thesis, University of Minnesota, St. Paul, MN.

Hawkins, L. L. 1986b. Nesting behavior of male and female whistling swans and implications of male incubation. *Wildfowl* 37:5–27.

Herbert, J. 1992b. Summary of Montana's tundra swan hunting seasons, 1970-90. 13th Trumpeter Swan Society Conference: 60-63.

Hobson, K. A. 2004. Preliminary stable isotope analysis of tundra swan feathers: A new technique for delineating breeding origins of wintering birds. 19th Trumpeter Swan Society Conference: 192-197.

Howie, R. R., and R. G. Bison. 2004. Wintering trumpeter and tundra swans in the southern interior of British Columbia. 19th Trumpeter Swan Society Conference: 16-28.

Huener, J. D. 1992. Tundra swan hunting in Utah. 13th Trumpeter Swan Society Conference: 76-80.

Johnson, M. A., and S. C. Kohn. 1991. Tundra swan hunting in North Dakota — results of the first season. 12th Trumpeter Swan Society Conference: 65-72.

Jordan, M. 1991. Trumpeter and tundra swan survey in western Washington and Oregon – January 1989. 12th Trumpeter Swan Society Conference: 14-17.

Kenow, K. P., J. M. Nissen, R. Drieslein, and E. M. Thorson. 2004. Tundra swan research needs on the upper Mississippi. 19th Trumpeter Swan Society Conference: 180-189.

Kohn, S. C., and M. A. Johnson. 1992. Results of tundra swan hunting seasons in North Dakota, 1988-90. 13th Trumpeter Swan Society Conference: 64-73.

Limpert, R. J., H. A. Allen, Jr., and W. J. L. Sladen. 1987. Weights and measurements of wintering tundra swans. *Wildfowl* 38: 108–113.

Limpert, R. J., and S. L. Earnst. 1994. Tundra swan (*Cygnus columbianus*). *The Birds of North America* 89. 30 pp. The Birds of North America, Inc. Philadelphia, PA: The Academy of Natural Sciences, and Washington, DC: American Ornithologists' Union. 20 pp.

Limpert, R. J., W. J. L. Sladen, and H. A. Allen, Jr. 1991. Winter distribution of tundra swans *Cygnus columbianus* breeding in Alaska and western Canadian Arctic. *International Swan Symposium* 3:78–83.

Loranger, A., and D. Lons. 1990. Relative abundance of sympatric trumpeter and tundra swan populations in west-central interior Alaska. 11th Trumpeter Swan Society Conference: 92-98.

Lumsden, H. G. 1984. The pre-settlement breeding distribution of trumpeter, *Cygnus buccinator*, and tundra swans, *C. columbianus columbianus,* in eastern Canada. *Canadian Field-Naturalist* 98:415–424.

Miller, S. L., M. A. Gregg, A. R. Kuritsubo, S. M. Combs, M. K. Murdock, J. A. Nilsson, B. R. Noon, and R. G. Botzler. 1988. Morphometric variation in tundra swans: Relationships among sex and age classes. *Condor* 90:802–815.

Monda, M. J. 1991. Reproductive ecology of tundra swans on the Arctic National Wildlife Refuge, Alaska. Ph.D. dissertation, University of Idaho, Moscow, ID.

Monda, M. J., J. T. Ratti, and T. R. McCabe. 1994a. Behavioral responses of nesting tundra swans to human disturbance and implications for nest predation on the Arctic National Wildlife Refuge (Abstract). 14th Trumpeter Swan Society Conference: 178.

Monda, M. J., J. T. Ratti, and T. R. McCabe. 1994b. Modification of tundra swan habitat by repeated use of nesting territories (Abstract). 14th Trumpeter Swan Society Conference: 179.

Nichols, J. D., J. Bart, R. J. Limpert, W. J. L. Sladen, and J. E. Hines. 1992. Annual survival rates of adult and immature Eastern Population tundra swans. *Journal of Wildlife Management* 56:485–494.

Pacific Flyway Council. 2001. *Pacific Flyway Management Plan for the Western Population of Tundra Swans*. Portland, OR: Pacific Flyway Study Committee, U.S. Fish and Wildlife Service.

Parmelee, D. F., and S. D. MacDonald. 1960. The birds of west-central Ellesmere Island and adjacent areas. *National Museums of Canada Bulletin,* 169: 1-101.

Parmelee, D. F., H. A. Stephens, and R. H. Schmidt. 1967. The birds of southeastern Victoria Island and adjacent small islands. *National Museums of Canada Bulletin,* 222: 1-229.

Paulin, D. G. 1996. Tundra swan use in California's Central Valley. 15th Trumpeter Swan Society Conference: 48- 52.

Paulin, D. G., and E. Kridler. 1988. Spring and fall migration of tundra swans dyed at Malheur National Wildlife Refuge, Oregon. *Murrelet* 69:1–9.

Petrie, S. A. and K. L. Wilcox. 2004. Migration chronology of eastern population tundra swans (Abstract). 19th Trumpeter Swan Society Conference: 179.

Petrie, S. A., S. S. Badzinski, and K. L. Wilcox. 2002. Population trends and habitat use of tundra swans staging at Long Point, Lake Erie. *Waterbirds* 25 (Special Publication 1): 143–149.

Rea, C., R. Ritchie, A. Stickney, and J. G. King. 2007. Multi-year monitoring program for tundra swans on the north Slope of Alaska. 20th Trumpeter Swan Society Conference: 136-139.

Retterer, T. E. 1992. Nevada's tundra swan hunting program. 13th Trumpeter Swan Society Conference: 56-59.

Ritchie, R. J., J. G. King, A. A. Stickney, B. A. Anderson, J. R. Rose, A. M. Wildman, and S. Hamilton. 2002. Population trends and productivity of tundra swans on the central Arctic Coastal Plain, northern Alaska, 1989–2000. *Waterbirds* 25:22–31.

Roy, V. 1996. Trumpeter and tundra swans: Their history and future at the Bear River Migratory Bird Refuge. 15th Trumpeter Swan Society Conference: 53-61.

Scott, D. 1977. Breeding behaviour of wild whistling swans. *Wildfowl* 28:101–106.

Serie, J. R., D. Luszcz, and R. V. Raftovich. 2002. Population trends, productivity, and harvest of Eastern Population tundra swans. *Waterbirds* 25 (Special Publication 1):32–36.

Sherwood, G. A. 1960. The whistling swan in the West with particular reference to Great Salt Lake Valley, Utah. *Condor* 62:370–377.

Sladen, W. J. L. 1991. Comments from a tundra swan researcher [on the draft position statement on tundra swan hunting]. 12th Trumpeter Swan Society Conference: 76-77.

Sladen, W. J. L., and W. W. Cochran. 1969. Studies of the whistling swan, 1967-*1969*. *North American Wildlife and Natural Resources Conference Transactions* 34:42-50.

Snyder, L. L. 1957. *Arctic Birds of Canada*. Toronto, ON: University of Toronto Press.

Spindler, M. A., and K. F. Hall. 1991. Local movements and habitat use of tundra or whistling swans *Cygnus columbianus* in the Kobuk-Selawik Lowlands of northwest Alaska. *Wildfowl* 42:17–32.

Stewart, D. B., and L. M. J. Bernier. 1989. Distribution, habitat, and productivity of tundra swans on Victoria Island, King William Island, and southwestern Boothia Peninsula, N.W.T. *Arctic* 42:333–338.

Stewart, R. E., and J. H. Manning. 1958. Distribution and ecology of whistling swans in the Chesapeake Bay region. *Auk* 75: 203-212.

Swystun, H. A., G. Kofinas, J. E. Hines and R. D. Dawson. 2004. Local observations of tundra swans (*Cygnus columbianus columbianus*) in the Mackenzie delta region, Northwest Territories, Canada (Abstract). 19th Trumpeter Swan Society Conference: 190.

Swystun, H. A., R. D. Dawson and J. E. Hines. 2004. Factors influencing reproductive success of tundra swans (*Cygnus columbianus columbianus*) (Abstract). 19th Trumpeter Swan Society Conference: 191

Tate, D. J. 1966, Morphometric age and sex variation in the whistling swan *(Olor columbianus)*. M.S. thesis, University of Nebraska-Lincoln, Lincoln, NE.

The Trumpeter Swan Society. 1991. The Trumpeter Swan Society position paper on tundra swan hunting – adopted January 1990. 12th Trumpeter Swan Society Conference: 83-84.

Trost, R. E. 1999. Trumpeter swan Rocky Mountain Population range expansion and tundra swan hunting: Is there a middle ground? 16th Trumpeter Swan Society Conference: 59-60.

Trost, R. E., D. Luszcz, T. C. Rothe, J. R. Serie, D. E. Sharp, and K. E. Gamble. 1999. Management and hunt plans for tundra swans. 16th Trumpeter Swan Society Conference: 103-108.

Vaa, S. J. 1992. South Dakota tundra swan season – 1990. 13th Trumpeter Swan Society Conference: 74-75.

Vaa, S. J., M. A. Johnson, and J. L. Hansen. 1999. An evaluation of tundra swan hunting in the Central Flyway and concerns about trumpeter restoration. 16th Trumpeter Swan Society Conference: 109-111.

Wilk, R. J. 1993. Observations on sympatric tundra, *Cygnus columbianus*, and trumpeter swans, *C. buccinator*, in north-central Alaska, 1989–1991. *Canadian Field-Naturalist* 107: 64–68.

Wilkins, K., R. Malecki, S. Sheaffer, and D. Luszcz. 2001. Eastern Population tundra swans: Population status, survival, and movements. *North American Swans* 30(1): 15-18.

Wilkins, K. A., R. A. Malecki, P. J. Sullivan, J. C. Fuller, J. P. Dunn, L. J. Hindman, G. R. Costanzo, and D. Luszcz. 2010. Migration routes and bird conservation regions used by Eastern Population tundra swans (*Cygnus columbianus columbianus)* in North America. *Wildfowl* 60:20–37.

Wilkins, K. A., R. A. Malecki, P. J. Sullivan, J. C. Fuller, J. P. Dunn, L. J. Hindman, G. R. Costanzo, S. A. Petrie, and D. Luszcz. 2010. Population structure of tundra swans wintering in eastern North America. *Journal of Wildlife Management* 74: 1107–1111.

TUNDRA (BEWICK'S) SWAN

Evans, M. E. 1976. Breeding behaviour of captive Bewick's swans. *Wildfowl* 26:117–130.

Evans, M. E. 1976. Aspects of the life cycle of the Bewick's swan, based on recognition of individuals at a wintering site. *Bird Study* 26:149–162.

Evans, M. E. 1980. The effects of experience and breeding status on the use of a wintering site by Bewick's swans *Cygnus columbianus bewickii*. *Ibis* 122:287–197.

Evans, M. E. 1982. Movements of Bewick's swans *Cygnus columbianus bewickii,* marked at Slimbridge, England, from 1960–1979. *Ardea* 70: 59–75.

Evans, M. E., and W. J. L. Sladen. 1980. A comparative analysis of the bill markings of whistling and Bewick's swans and out-of-range occurrences of the two taxa. *Auk* 97:697–703.

Scott, D. K. 1978. Social behaviour of wintering Bewick's swans. Ph.D. dissertation, University of Cambridge, Cambridge, UK.

Scott, D. K. 1981. Geographical variation in the bill patterns of Bewick's s wans. *Wildfowl* 32: 123-128.

Scott, D. 1967. The Bewick's swans at Slimbridge, 1966-67. *Wildfowl Trust Annual Report* 18: 24-27.

Scott, D. 1967. The Bewick's swans at Slimbridge, 1968-69. *Wildfowl Trust Annual Report* 20: 157-160.

Scott, P. 1966. The Bewick's swans at Slimbridge. *Wildfowl Trust Annual Report* 17: 20-26.

COSCOROBA

Blake., E. R.1977. *Manual of Neotropical Birds*. Vol. 1. Chicago, IL: University of Chicago Press. 674 pp.

Gibson, E. 1920. Further ornithological notes from the neighbourhood of Cape San Antonio, Buenos Aires. *Ibis* 11: 1-97.

Johnson, A. W. 1965. *The Birds of Chile, and Adjacent Regions of Argentina, Bolivia and Peru*. Vol. 1. Buenos Aires, Argentina: Platt Establecemientos Graficos. 397 pp.

Weller, M. W. 1975. Habitat selection by waterfowl of Argentine Isla Grande. *Wilson Bulletin* 87: 83-90.

Woods, R. W. 1975. *The Birds of the Falkland Islands*. Oswestry, U.K.: Anthony Nelson.

Zea Books by Paul A. Johnsgard

Swans: Their Biology and Natural History (2016)

Birding Nebraska's Central Platte Valley and Rainwater Basin (2015)

At Home and at Large in the Great Plains: Essays and Memories (2015)

Global Warming and Population Responses among Great Plains Birds (2015)

Música de las Grullas: Una historia natural de las grullas de América,
trans. Enrique H. Weir & Karine Gil-Weir (2014)

Birds and Birding in Wyoming's Bighorn Mountains Region (2013)
by Jacqueline Canterbury, Paul Johnsgard, & Helen Downing

Birds of the Central Platte River Valley and Adjacent Counties (2013)
by Mary Bomberger Brown & Paul A. Johnsgard

The Birds of Nebraska: Revised Edition, 2013 (2013)

A Prairie's Not Scary (2012)

Wings over the Great Plains: Bird Migrations in the Central Flyway (2012)

Wetland Birds of the Central Plains: South Dakota, Nebraska and Kansas (2012)

Rocky Mountain Birds (2011)

A Nebraska Bird-Finding Guide (2011)

Printed in Great Britain
by Amazon